Robotics

A User-Friendly Introduction

Robotics
A User-Friendly Introduction

Ernest L. Hall
Bettie C. Hall

University of Cincinnati

Holt, Rinehart and Winston
New York · Chicago · San Francisco · Philadelphia
Montreal · Toronto · London · Sydney · Tokyo
Mexico City · Rio de Janeiro · Madrid

Front cover photograph courtesy of Computer Graphics Lab, New York Institute of Technology.
Back cover photographs courtesy of Cincinnati Milacron, DeVilbiss, and ASEA.

Copyright © 1985 CBS College Publishing
All rights reserved.
Address correspondence to:
383 Madison Avenue, New York, NY 10017

Library of Congress Cataloging in Publication Data

Hall, Ernest L.
   Robotics, a user-friendly introduction.

   Bibliography: p.
   Includes index.
   1. Robotics.   I. Hall, Bettie C.   II. Title.
TJ211.H24   1985   629.8'92   85-785

ISBN 0-03-069718-2

Printed in the United States of America

Published simultaneously in Canada

5 6 7 8   016   9 8 7 6 5 4 3 2 1

CBS COLLEGE PUBLISHING
Holt, Rinehart and Winston
The Dryden Press
Saunders College Publishing

For our students

Introduction

Robotics: A User-Friendly Introduction is an introductory text designed for undergraduate robotics courses. Written in a style that is a careful mixture of the vernacular and pedagogy, the book is both readable and informative. A complete coverage of the emerging field of robotics is presented, including a description of the components and operation of industrial robots, intelligent robot programming, sensors for intelligent robots, industrial robot applications, and future robot considerations. A chapter on the history of robotics is also included to provide a means of discovering the origins of this fascinating machine. Economic considerations and methods for the justification of an industrial robot purchase are also presented. The social and political implications and impacts of robots are also presented in two thought-provoking chapters.

Although the overall book is introductory, some previous computer-programming experience is helpful for portions of the material. The text is recommended for a first course in industrial robotics.

Contents

Introduction vii
Preface xiii
To the Instructor xv

1. An Introduction to Robotics 1
    1.1   What Is a Robot? *2*
    1.2   Why Robots Are Needed *4*
    1.3   Overview of Robot Applications *5*
    1.4   Chapter Previews *7*

2. History of Robotics 9
    2.1   Universal Machines *9*
    2.2   The Industrial Revolution *10*
    2.3   Prosthetic Devices *11*
    2.4   Remote Manipulators *12*
    2.5   Key Events in the Recent History of Robotics *18*

3. Robot Components and Operation 24
    3.1   Basic Components *24*
            Manipulators *24*
            Wrists *32*
            End Effectors *36*
            Control Units *44*
            Power Units *44*
    3.2   Operation of Industrial Robots *47*
            Nonservo Robot Operation *52*
            Servo Operation *54*

Contents

- 3.3 Methods of Motion Control 56
  - Continuous Path Motion 56
  - Point-to-Point Operation 56
  - Joint Interpolated Motion 58
  - Controlled Path Motion 58
- 3.4 Hierarchy of Control for Robots 60
- 3.5 Line Tracking with Industrial Robots 61
- 3.6 Modular Robots 63

4. Intelligent Robot Programming   66

- 4.1 Artificial Intelligence 67
- 4.2 Machine Intelligence 73
  - Robot Checkers Player 75
  - A Robot Solution to Rubik's Cube 75
- 4.3 Voice Control of Robots 79
- 4.4 Programming a Robot 80
  - Level 1 Robot Languages 83
  - Teach-by-Example Programming 94
  - Off-Line Programming 104

5. Robot Sensors   106

- 5.1 Robot Sensor Classification 108
- 5.2 Image Processing for Robot Vision 110
- 5.3 Robot Vision Specification 112
  - Illumination Systems 113
  - Camera Positioning, Focus, Zoom, and Aperature Control 113
  - Camera Selection 113
  - Digitizing an Image 116
  - Processing Examples 118
- 5.4 Commercial Robot Vision Systems 123
  - Silhouette Robot Vision 123
  - Gray-Scale Robot Vision Systems 126
  - Three-Dimensional Robot Vision 126
- 5.5 Range and Proximity Sensors 127
- 5.6 Tactile Sensors 129
- 5.7 Sensors for Mobile Robots 133
- 5.8 Sensor and Control Integration 138

6. Applications of Industrial Robots   139

- 6.1 Basic Industries Involved in Manufacturing 139
- 6.2 Fundamental Operations in Material Processing and Assembly 146

Machining   *146*
Processing of Raw Materials   *147*
Casting Process   *148*
Heat Treatment   *149*
Welding Applications   *150*
Assembly Operations   *155*
Finishing Applications   *156*

6.3   Materials Handling and Storage Applications   *160*
Machine Loading and Unloading   *160*
Palletizing   *163*
Bin Picking   *165*

7. Economic Considerations and Justification of Industrial Robots   168

7.1   Economic Considerations   *168*
7.2   Economic Justification of a Robot   *175*
7.3   Economic Justification of a Work Cell   *182*
7.4   Economic Considerations for the Automated Factory   *185*

8. Social Impact of Industrial Robots   *187*

8.1   New Requirements in Education and Training   *188*
8.2   Effects of Robots on Employment   *191*
8.3   Effects of Robots on Workers   *193*
Underemployment and Job Satisfaction   *194*
Robots and Organized Labor   *199*
8.4   Resistance to Change   *199*

9. Responsible Technology   203

9.1   Economic and Political Impacts of Robots   *203*
9.2   Robot Development in Other Nations   *205*
9.3   Growth of the Robotics Market in the United States   *208*
9.4   Anticipated Benefits   *210*
9.5   Alternatives   *212*
9.6   Conclusions   *213*

10. Robots Today and Tomorrow   214

10.1   Robot Developments   *214*
Automated Factories   *214*
The Need for Standardization   *214*
10.2   Ongoing Research and Future Applications   *215*

　　　　　　　Advances in Robot Intelligence　*215*
　　　　　　　The Military　*216*
　　　　　　　Domestic and Entertainment Robots　*216*
　　　　　　　Robots in Medical and Patient Care Applications　*220*
　　　　　　　Robots in Hostile or Remote Environments　*221*
　　　　　　　Robots in Education　*221*
　　　　　　　Robots in Agriculture　*223*
　　　　　　　Mobile Robots　*226*
　　　10.3　Conclusion　*227*

Bibliography　　　　　　　　　　　　　　　　　　　　　　　**229**

Glossary　　　　　　　　　　　　　　　　　　　　　　　　**235**

Index　　　　　　　　　　　　　　　　　　　　　　　　　**249**

Preface

The purpose of this book is to provide a user-friendly introduction to the important and exciting field of robotics. Robots will certainly affect our lives in important ways in coming years—economically, socially, and politically. They are already changing the way we do much of our work in factories and promise to change the way we work in our home and offices in the future. They are beginning to have an impact on the productivity of our industries and on the competitiveness of our products in the world market.

Robots are assuming many of the dangerous and difficult tasks that humans now do; this is one of the greatest benefits offered by robots. As robots assume tasks in hazardous environments, perhaps all industrial diseases that now afflict human workers can be eliminated. Another benefit brought about by the use of robots concerns what sort of work we deem fit for humans to perform. Robots are able to do simple, monotonous, repetitive work much better than humans because they do not get bored or tired. Humans were surely never intended for such work, or we would not have been equipped with the wonderful assortment of sensory, intellectual, and perceptual capabilities that goes into our makeup. Robots, on the other hand, are designed and intended for these jobs, freeing humans for more challenging and satisfying kinds of work.

This book is directed toward those readers who are willing to face the challenges of robotics, who are concerned about the impacts of modern automation on society, and who may someday live in a world with millions of robots. Through explanations, examples, drawings, and photographs, the authors hope to give readers a general idea of what robots are, how they work, what they can and cannot do, how people react to them, how they affect us now, and how they might affect us in the future. Although a good deal of attention is given to the industrial robot (because it is the most numerous and well-known type used today), we are also concerned with the use and impact of intelligent robots. These wonderful combinations of machine and computer powers have the potential and capability to make our best dreams of science fiction come true.

At present, most of the educational literature on robotics has been written in the highly precise and technical language of engineering mathematics. This would be fine if no one but engineers needed to know about robots. However, the impact of these universal machines has created a need for an interdisciplinary, general introduction to robotics. It is this need that has motivated the writing of this book.

We gratefully acknowledge the assistance and encouragement provided us by our friends at the University of Cincinnati, the University of Tennessee, Cincinnati Milacron, the Oak Ridge National Laboratory, and the other robotics educational and research facilities who assisted us in the collection of information for this book.

Our special thanks go to Professors Ivan Morse, Ronald Huston, and Richard Shell, and to Dean Lewis Laushy at the University of Cincinnati for providing their encouragement and support of this work. We also deeply appreciate the assistance of James Geier, Richard Messinger, James Gavin, Alfred Scheide, Dick Carrico, Mertin Corwin, and Richard Hohn of Cincinnati Milacron for sharing their robotics expertise. Our warmest thanks and sincere appreciation go to Ronald Tarvin of Cincinnati Milacron, who freely shared his robotics course notes, provided suggestions, and reviewed the manuscript. Finally, we would like to thank Professors June Adamson and Michael Keene of the University of Tennessee for their inspiration and high standards of clear communication, without whose encouragement this book would never have been attempted.

We would also like to thank our reviewers, whose keen observations and practical criticisms helped us immensely in seeing the book through others' eyes. They caught several errors and noted missing details; any faults of the book that remain are the sole responsibility of the authors.

We would also like to note that our editor, Mr. John Beck, has been a constant guide and source of inspiration to us throughout the preparation of this work, and to him we extend our sincere appreciation.

Ernest L. Hall
Bettie C. Hall

To the Instructor

This book has been designed for use in an introductory robotics course for a general audience. The material has been developed from a series of robotics courses taught at the University of Cincinnati and the University of Tennessee. As you will note, a careful mixture of vernacular and pedagogic material is used for clarity and ease of understanding. Also, a fair sprinkling of opinion and conjecture is included to convey to students an understanding of some of the paradox, conflict, and problems that need to be thought about but are rarely mentioned in standard robotics texts.

As an introductory technology course at the freshman or sophomore level, you may wish to concentrate on Chapters 1, 2, 4, and 5. Chapter 2 provides an opportunity to discuss the history and evolution of robots and can lead to an understanding of the sophistication of the modern, industrial robot. The book assumes some knowledge of programming, especially in Chapter 4.

For an introductory awareness course at this level, the entire text may be assigned for reading and special topics selected for class discussion from either the material or problems.

As an engineering text for juniors, seniors, and graduate engineering and computer science students for a one-quarter course, the material in Chapters 1, 2, 3, 6, and 7 can be covered thoroughly in about 30 hours of lecture. A second quarter could be based on Chapters 4, 5, 8, 9, and 10, with an emphasis on the use of intelligent robots and the intelligent use of robots. The material could also be used in a three-quarter course, especially if projects were assigned each quarter. For example, in the first quarter, a project on the economic justification of a robot, such as that given in Chapter 7, could be assigned. In the second, a project involving robots and sensor systems could be used as the basis of an intelligent robot project. For the third quarter, a robot design or major application project could be assigned or studied.

For a two-semester course, Chapters 1, 2, 3, 4, and 7 could be covered in the first semester. A project on the economic justification of an industrial robot would be appropriate. During the second semester, Chapters 5, 6, 8, 9, and 10 could be covered, with emphasis on intelligent robots. Robot programming and simulation could also be emphasized.

Supplementary materials, such as videotapes, are available from Robotics International, through either the local chapter or from headquarters which is located at 1 SME Drive, Dearborn, Michigan. We have found most manufacturers very willing to assist with plant tours of local facilities. A well-equipped robotics laboratory is very helpful. A set of slides of the illustrations in this book is available from the authors.

Robotics
A User-Friendly Introduction

An Introduction to Robotics

> The discovery of nature, of the ways of planets, plants and animals, required first the conquest of common sense. Science would advance, not by authenticating everyday experience, but by grasping paradox, adventuring into the unknown. . . . Nothing could be more obvious than that the earth is stable and unmoving, and that we are the center of the universe.
>
> D. J. Boorstin (1983)

Robotics is the science of designing, building, and applying robots. In this chapter, we will explain what robots are and why we need to understand them. We could just assume that everyone understands what a robot is, but many of our ideas about what robots are and what they can do may owe their existence to fiction, to stories and movies. For example, the word *robot* is based on the Czeck word for slave, and was introduced into our culture in the early 1920s in a play by Karel Čapek about mechanical men that rebel against their human masters. The word *robotics* was coined by the renowned science fiction writer, Isaac Asimov, in the 1942 science fiction story, "Runabout" (Asimov, 1982). However, it has only been in the past 25 years that real robots and the serious study of them has advanced from the realms of fiction to the laboratories of universities and into the factories of industry.

Today, robotics has developed into a solid discipline of study that incorporates the background, knowledge, and creativity of mechanical, electrical, computer, industrial, and manufacturing engineering. Even without the fictional trappings of supercapabilities, robotics is an exciting, challenging field of study. Those who are students today will be the robotics experts, designers, programmers, users, and teachers of tomorrow. Today's robotics students will be very much a part of the evolution of robots toward their ultimate capabilities. Students in every discipline from engineering to sociology have the opportunity to become involved with what promises to be the greatest innovation in our means of performing work since the computer.

However, as we study robotics, it is prudent to remember that perhaps many of our assumptions about what robots are and what they can or cannot do are founded on fiction. In adventure stories, robots are able to navigate spaceships, care for children, and vie with humans for control of the galaxy. In the real world, robots do such practical work as painting and welding cars or handling radioactive elements. Although the fictitious robots may not necessarily be impossible, most experts would agree that it is far easier to imagine a mechanical marvel than it is to build one that works. Besides, we don't need or want robots that fight us. We only need robots that help us do our work faster or more efficiently so that we humans can enjoy a better life. It is important that we learn about robots and explore their capabilities so that ours never becomes a society in which the human serves the machine, but one in which the machine always serves humans.

1.1 What Is a Robot?

If you walk into the Nissan truck factory in New Smyrna, Tennessee, expecting to see shiny androids like C3PO assembling parts, you will be severely disappointed. The modern, industrial robot has far more in common with an ordinary piece of machinery than with a human. This only makes sense, since robots are machines. An example of a robot at work is shown in Figure 1–1. This industrial robot is trimming plastic dashboard components by moving them under a high-power laser. What is the difference between this robot and any other piece of automated machinery? Why is this machine called a "robot," not just an "automatic dashboard trimmer?" The robot is a special kind of automated machine. A robot can do not only this particular job of trimming, but it can be programmed and retooled to do many different jobs. This programmability and versatility is why all robots are automated machines, but all automated machines are not robots.

There is only one definition of an industrial robot that is internationally accepted. It was developed by a group of industrial scientists from the Robotics Industries Association (formerly the Robotics Institute of America) in 1979. They defined the industrial robot as ". . . a reprogrammable, multifunctional manipulator designed to move material, parts, tools, or specialized devices through various programmed motions for the performance of a variety of tasks." Let's take a close look at this definition to see just what it implies.

The first key word is *reprogrammable*. This implies that a robot is a machine that cannot only be programmed once, but can be programmed as many times as one likes. Many electronic devices we use every day contain computer chips that are programmable. Programs are written on the chips of digital watches, for instance, that instruct them to do such things as play "Dixie" as an alarm. These programs cannot be easily changed, however. There is no allowance for input by the owner. You cannot, for example, put a song of your own into the watch when you get tired of waking up to

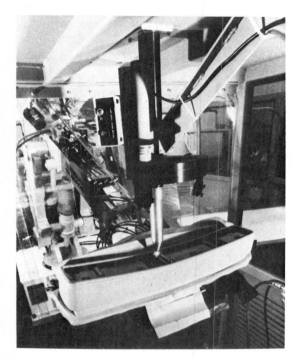

Figure 1–1. A Cincinnati Milacron T3 746 robot trims plastic dashboard components for Ford by inserting them in a shielded booth and maneuvering them under a CO_2 laser. After trimming, parts are placed on an indexing output conveyor at left. The laser is a 500-watt Versa Lase V500 made by Photon Sources, Livonia, Michigan. (Courtesy of Cincinnati Milacron.)

"Dixie." The programs are "burned in" by the manufacturer. A robot, however, contains a program that is accessible, that can be changed, added to, or deleted, as the user chooses. A robot can have many programs to do different things in any sequence whatever. And, of course, to be programmable, a robot must have a computer that can be fed new instructions and information. The computer can be either "on board," which means the computer console is mounted on the robot itself, or it can be "remote," which means the computer that controls the robot can be anywhere you like as long as it can communicate with the robot.

The next key word in the RIA definition is *multifunctional,* which implies that the robot is versatile, that is, can perform more than one task. The same industrial robot used for laser cutting in Figure 1–1 could, with a simple change of end tooling, also perform welding, painting, or assembly operations.

The third key word is *manipulator,* which implies that a robot has a mechanism of some sort for moving objects for the performance of its work. It's the manipulator that separates a robot from a computer, just as it's the reprogrammability and versatility that separate a robot from other kinds of automated machines.

Finally, let's consider the meaning of *various programmed motions*. This implies that the robot is dynamic; that is, it is characterized by continuous, productive activity.

Although this definition may seem very broad and somewhat ambiguous, it does serve to separate industrial robots from, for instance, fixed-sequence automated machinery, or from multifunctional machines, such as food processors, that are equipped with interchangeable parts to perform various tasks, from blending sauces to grinding beef. It also removes robots far from the realms of science fiction, since any anthropomorphic (humanlike) characteristics a robot may or may not possess are merely a matter of efficacy.

From this perspective, then, the robot can be considered a major advance in the logical progression in the development of automated machines. We have moved from building machines that can do one job with human control to machines that can do many different jobs without any human control. The first industrial revolution has been said to be the start of an era of general industrial use of power-driven machines. The modern industrial renaissance may be called an era in which we are building machines capable not only of building other machines, but also of repairing and "reproducing" themselves. Some of the latest research is in equipping robots with such sensory apparatus as "eyes" and artificial intelligence sufficient to allow the robot to "learn" or to adapt to variable conditions in its working environment. Good questions at this point might be, "But why do we need such sophisticated machines, or machines that can do several different jobs? Aren't the automated machines we now have sufficient?"

1.2 Why Robots Are Needed

The basic reason for using robots rather than other automated machinery to produce goods is closely related to their versatility and programmability. This versatility can be translated into increased productivity, improved product quality, and decreased production costs in several ways. In a market with periodic product changes, the cost of reprogramming and retooling a robot is much less than the cost of retooling a fixed automation machine. If product changes are brought about by inflation or competition, again, the versatility of the robot is important in permitting minor product changes quickly. Because robots assume many dangerous or annoying jobs, many employee injuries are eliminated that comprise a very costly element of production. Because robots perform their jobs the same way every time, they produce consistent quality in their goods, which provides the manufacturer with definite advantages. Predictable production rates permit better inventory control. Each savings in the cost of the value added to a product results in improved competitiveness in all markets. One other advantage that robots offer is that they can be used to make small batches of products, but hard automation is generally efficient only in mass, standardized production.

There are many other reasons for using robots. The cost of human labor has been rising at a marked rate for the past several years, but the speed at which humans can work has not increased. Human labor costs are now so high that the cost of using machine labor is usually much more attractive. Although the machine may have a high initial cost, it can increase production by working faster and for more hours per day. Machines have improved the working conditions of humans by assuming hazardous and monotonous jobs and reduced production costs because they produce fewer "rejects" that humans sometimes produce through fatigue or boredom. Robots improve productivity in a variety of applications from processing raw materials to assembling automobiles. They are especially useful for work in hostile or dangerous environments, such as in outer space or on the ocean floor. Finally, robots are fun to work with. They provide challenging opportunities to everyone from hobbyists to the most advanced robotic designers. The robot is not only going to be the servant we have always dreamed of owning, but also the ultimate machine.

1.3 Overview of Robot Applications

Industrial robots have been used in a wide variety of manufacturing applications. Hot, dirty, dangerous foundry work in which molten metal is poured into castings was one of the first jobs in which robots were successfully used. Welding operations, in which consistency of the spot or seam weld is essential but which also produces a hot, ozone atmosphere annoying or hazardous to humans, has become another widely used application. Hazardous spray painting is another application in which robots are important, because robots can safely apply extremely thin coats of paint consistently, which significantly reduces the amount of paint needed per part. Back-breaking, dangerous, and tedious machine loading and unloading is another task to which robots are often applied. An example of a machine-loading robot is shown in Figure 1–2. Such robots are often the central element of an automated work cell, which is a coordinated collection of machines designed to perform a set of tasks, such as machining parts or spray painting. Assembly of automobiles, electric motors, computers, and even robots are newly proven areas of robot application.

Most robots used in these applications are deaf, dumb, blind, and stationary. Thus, these robots are not used so differently from other kinds of automated machines. However, an entirely new phase in robotics applications has been opened with the development of "intelligent" robots. An intelligent robot is basically one that is equipped with some sort of sensory apparatus that enables it to sense and respond to variables in its environment. Much of the research in robotics has been and is still concerned with how to equip robots with seeing "eyes" and tactile "fingers." Artificial intelligence that will enable the robot to respond, adapt, reason, and make decisions in reaction to changes in the robot's environment are also inherent capabilities of the intelligent robot. For example, one of the most important considerations in using a robot in a workplace is safety. If a robot could be equipped with sensory apparatus that detect the presence of humans, it could be programmed to automatically shut down its operation if it sensed the presence of a human within its work envelope. Intelligent robots that have already found successful application can do such things as "see," "hear," and "feel." The development

Figure 1–2. The Prab Model 5800 nonservo controlled industrial robot shown loading cylinder heads in a machining application. (Courtesy of Prab Robots, Inc., Kalamazoo, Michigan.)

of sensors coupled with recent innovations in robot mobility have enabled robots to move out of factories and into such varied environments as orange groves, sheep farms, and hospitals. Robots are also used in domestic and entertainment applications. Some futurists, such as Bill Bakaleinikoff of Superior Robotics of America, see the domestic robot as a mobile entertainment or sentry machine. Others see them as personal slaves that fetch and carry for their owners. Such applications are in their earliest stages now, but offer an exciting challenge to future roboticists. We can expect many new industries to develop around the creation and applications of robots. The potential applications of these new robots seem limited only by human imagination and creativity.

Does this mean robots will take over all work? Will there be nothing left for people to do? They may take over all the hardest work, all the dangerous work, and all the boring work, because, as consistent workers, robots are vastly superior to humans. They can run 24 hours a day, 7 days a week, year after year without stopping. Robots as imitators of human beings are vastly inferior, however. No robot has ever been, and is not likely ever to be, invented that can do everything people can do. People as workers are amazingly adaptive and creative. We are capable of learning thousands of jobs in just one lifetime. Furthermore, we have a marvelous assortment of senses coupled with the most sophisticated intelligence system in existence. We cannot even begin to assess the total or potential abilities of a human being. Finally, we have feelings, emotions, and biologic responses that make us uniquely suited to helping other people. For example, contrary to some popular notions, robots would probably not be good baby-sitters, because young children need the kind human responses of which machines are incapable. The simplest answer to the question of whether robots will take over all work can be seen by considering all the ways people can help other people. Obviously, we will never run out of things for people—or robots, for that matter—to do.

1.4 Chapter Previews

In Chapter 1, we have begun with a general introduction to robotics. Some perspectives concerning robots have been reviewed, and some of the key terms have been defined. More complex terms are defined in the Glossary at the end of the book. What robots are, why we need to use them, and an overview of applications were also presented.

In Chapter 2, we present a brief history of the development of robots, showing how the confluence of automated machines, remote manipulators, and prosthetics has led to the modern, industrial robot. Some key events in the history are listed in chronological order, and early robots and their applications are also discussed.

In Chapter 3, the structure of an industrial robot is detailed to provide the reader with a basis for understanding robots in general—why they look and work the way they do. The basic components of industrial robots, such as wrists, end effectors, control units, and power sources, are also described. The operation and methods of motion control of robots are also discussed.

Chapter 4 is concerned with intelligent robot programming. Several examples of intelligent robots are described, as well as the characteristics of intelligent robots, one of which is their ability to adapt to and perhaps control their environments. Robot programming is introduced, and examples of robot programming languages are considered in detail.

The types of sensors available for use with robots are considered in Chapter 5. Vision, tactile, temperature, proximity, and other sensors used separately or integrated together are described. Sensors for mobile robots are briefly considered. The problem of control integration is also considered.

Chapter 6 presents applications of robots today in manufacturing tasks, such as welding, painting, materials handling, and assembly. The basic industries involved in manufacturing and the fundamental operations are described. Single-robot applications and manufacturing systems are also discussed, especially where the robot is the central element in these factory-of-the-future arrangements, moving parts from one location to another, loading and unloading machines, and performing production or assembly operations.

Chapter 7 centers on the economic justification for robots. Questions from both management and labor that must first be considered are discussed. From the management point of view, cost is used to justify the use of a flexible manufacturing cell and an

8 | An Introduction to Robotics

automated factory of the future. Costs and benefits resulting from robotic installations are also discussed.

Chapter 8 deals with the social aspects of robots. The findings of sociologic impact studies are reviewed, and suggestions are given on how to deal with the social consequences and human-machine interrelations that must result from the implementation of large numbers of robots in the work force.

Chapter 9 reviews the various political implications of what many people see as an "invasion" of thousands of robots into our society. There are people who fear robots will spark massive unemployment or inflation and who want strict controls on robot development. This chapter tells why most of these fears are unnecessary and discusses the role of the United States as a manufacturing nation, international economic competition, and the cooperation required for maintaining our present way of life.

Chapter 10 tells where robotics is headed in light of ongoing research. It describes those areas where much remains to be done, what is possible, and what is probable in our immediate future. The need for standardization in the robotics community is discussed, as well as new and unusual applications of mobile and multisensory robots. The book concludes with a discussion of how technology can best be assimilated into a society.

Questions

1. Where do the words *robot* and *robotics* come from?
2. What is an industrial robot, and what makes it different from other automated machines?
3. List five advantages robots offer over other automated machines.
4. How can robots allow humans to be more productive?
5. How do you think the modern industrial renaissance differs from the industrial revolution?

History of Robotics

> Precisely because the clock did not start as a practical tool shaped for a single purpose, it was destined to be the mother of machines. . . . The enduring legacy of the pioneer clockmakers, though nothing could have been further from their minds, was the basic technology of machine tools.
>
> D. J. Boorstin (1983, p. 64)

2.1 Universal Machines

The clock was the first universal machine. Its invention and development influenced many discoveries, such as the prediction of seasons for agriculture, the measurement of longitude for navigational exploration and discovery, the measurement of heart rates, and the coordination of military maneuvers. The development and significance of the "mother of machines" is traced by Boorstin from water clocks, to clocks that called people to prayers, to the eighteenth century androids (Boorstin, 1983).

The automobile is also a universal machine. The gasoline engine is totally integrated into modern civilization. The digital computer is another example of a universal machine. Its contributions to planetary exploration, industry, medicine, the military, and, more recently, domestic applications are well known.

The industrial robot is our most recent development of a universal machine and ranks in importance with the clock, automobile, and computer. Its applications in industry are well known; however, the expansion to agricultural, space and sea exploration, medical, and domestic applications has already begun.

To fully appreciate the capabilities of the modern industrial robot, it is useful to trace the confluence of mechanical, electrical, and industrial technologies that led to its development. This is a difficult task. In this chapter, we present a selected, brief introduction to these developments. We do this so that we may not only understand how robotics got where it is today, but might also perceive where the field is going tomorrow.

The modern robot has existed only since the 1950s, but the idea of a device that can move about under its own power and control has been important to humankind for millenia. For example, the Egyptians in 3000 B.C. built water clocks and water-powered, jointed, moving figures. The ancient Greeks also had life-sized, automatic puppets that were used in dramatic productions (Boorstin, 1983, p. 93; Malone, 1978, p. 24). The Chinese and Ethiopians also built devices that performed interesting or amusing sequences of motions (Albus, 1981, p. 229). From the fourteenth through the eighteenth centuries, many mechanical devices were built that were very lifelike and performed some realistic actions. These automata culminated in the fascinating automatons of Jacquet-Droz and Maillardet (Geduld and Gottesman, 1978, p. 22).

The principles of the wheel, lever, winch, screw, and windlass with rope and pulley, powered by various sources, such as water, steam, combustion, or human or animal effort, were used to develop all automated machines. Each new machine and power source represented another step toward modern robots.

2.2 The Industrial Revolution

The basic developments that led to the modern industrial robot were started during the industrial revolution, when automatic power sources, machined parts, and controllers were developed. The industrial age was ushered into our work by machines that used steam power to perform tasks. The general use of this power source led to the design of a new class of automated machines. For example, in the textile industry, new machines could produce goods much faster than could humans using traditional, hand methods. With the invention of the steam engine by James Watt in 1769, Edmund Cartwright's power loom in 1785, and Eli Whitney's cotton gin, automation was here to stay.

The next step in the development of the robot was in machines that could build machine parts. In 1800, the metal lathe was invented by Henry Maudslay. His lathe design, in which the part to be shaped is rotated about a horizontal axis and shaped by a fixed tool, is still used today for metal working. In 1818, Eli Whitney invented the milling machine. In this milling machine, the cutting tool is rotated against the workpiece, turning out uniform parts. He, Samuel Colt, and other gun manufacturers used these milling machines and lathes to produce standard, interchangeable, precision parts that were assembled on a mass-production line (*Science and Invention Encyclopedia*, Vol. 2, p. 162).

The final step toward robots was the development of controllers. In 1805, Joseph Marie Jacquard perfected a punched-card control mechanism for his automatic looms. His design is considered to be the main precursor of the stored program computer. Some 10 years later, Charles Babbage, who had successfully built a six-digit adding machine, started the construction of his difference engine to produce the trigonometric, logarithmic, and other mathematical tables needed in navigation and mathematics.

Although the general-purpose robot had still not yet been conceived, a motorized rotary crane with a gripper was designed and patented by Steward S. Babbitt in 1892 for removing hot ingots from furnaces (U.S. patent 484,870). Then, in the 1930s, a device for spray painting was invented. This device permitted the operator to move the painting apparatus through a series of motions while the signals were recorded on magnetic storage media. These signals could then be played back to control the painting motion. Inventions of this type were patented by Willard L. V. Pollard (U.S. patent 2,286,571;1942) and Harold A. Roselund (U.S. patent 2,344,108;1944), and may be considered precursors of the modern painting robot.

2.3 Prosthetic Devices

Studies and work in prosthetics were also important to the development of the modern industrial robot. In designing artificial limbs, scientists had to study the human mechanisms that made our movements and abilities possible, which led machine designers to think about how the human arm, hand, and fingers function together to do work. An early example of a prosthetic device was described by Herodotus in 500 B.C. He told of a captive who severed his own foot at the instep to free himself from his bonds, then designed a wooden foot to replace it. In the Second Punic Wars, about 218 B.C., a Roman general named Marcus Sergius lost his right hand. He had an iron replacement made that reportedly served him quite well in battle. War injuries provided an impetus to develop more sophisticated types of prosthetics to replace severed limbs. During the 1500s, Götz von Berlichingen lost his hand in battle and replaced it with a mechanical hand with movable fingers.

Prosthetic arms were developed in the sixteenth century. These were important developments toward transmission and control mechanisms that permitted the movement of mechanical fingers and thumbs so that the user could perform simple tasks, like grasping and holding. In this century, electrically powered artificial hands were developed. These were powered by electrical impulses picked up from nerve endings in the wearer's severed limb and magnified and transmitted to the hand to power its movements. These worked well until the thalidomide disaster of the 1960s. Thalidomide victims were born with stunted limbs and had no nerve endings to which electrodes in the standard prostheses could be attached. This led to the development of artificial limbs that could be operated pneumatically through compressed gas connected to pistons inside the artificial arm. These devices responded to the bulge or hardness of a contracting muscle. One of the latest developments has been the myoelectric arm. This type of device can be operated either by electric motors or compressed gas, and again pick up the electrical impulses sent to muscles to activate the device. Wire electrodes are inserted into the muscle to pick up these electrical potentials, and when the muscle is signaled to contract, these electrodes pick up the signal and amplify it, so that patients can use the same impulses to control an artificial hand as they would a real one. A modern artificial hand is shown in Figure 2–1.

The development of artificial limbs contributed much to robotics because they provided studies in transmission, manipulation, and control systems. The same designs used to construct artificial arms and hands were later used to design remote manipulators and, subsequently, the robot manipulator. Conversely, the study of robotics is now leading to further developments in prosthetics.

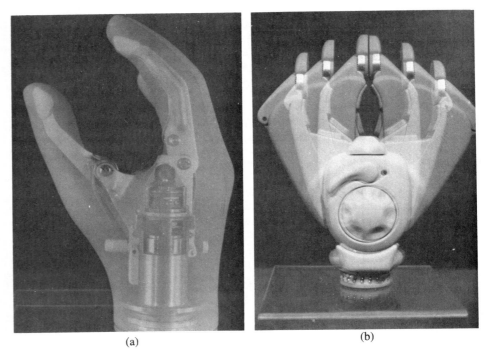

(a) (b)

Figure 2–1. A modern prosthetic hand. (a) This x-ray image of the device shows the outer covering, the internal mechanical structure, and the servo motors. (b) The prosthetic hand's gripping action is shown in this multiple-exposure x-ray image. (Courtesy of Otto Bock Orthopedic Industry, Inc., Minneapolis, Minnesota.)

2.4 Remote Manipulators

John Naisbitt states in his book, *Megatrends,* that "robots for dangerous tasks and toys were following the path of least resistance" (Naisbitt, 1982). One of the first areas of remote manipulator applications was in the handling of radioactive materials. During the 1940s, "hot cells," which were protected rooms containing radioactive elements, were set up for research. Radioactive materials could be safely stored and transported in lead containers called "pigs," but removing the material for use in the hot cell was a problem. Some method for manipulating this material without direct human contact was necessary due to its toxic nature. This led to the development of the "master-slave" manipulator. In this system, the "slave" mechanical arm inside the radioactive environment mimics the motions made by the "master" arm outside the hot cell.

The first master-slave manipulator was developed by Ray C. Goertz and others at the Argonne National Laboratory in 1944. In this manipulator, the master arm outside the hot cell was linked mechanically to the slave arm inside the hot cell. The human operator directed the motion of the master arm, which in turn directed the motion of the slave arm. However, these mechanical linkages often provided awkward or difficult manipulation of the slave arm because the operator could not feel the collision of the slave arm with obstacles or objects. Goertz described the operations of his manipulator as a series of collisions, as follows: "In all these operations, the manipulator must come into physical contact with the object before the desired force and movements can be made on it. A collision occurs when the manipulator makes this contact. General purpose manipulation consists essentially of a series of collisions with unwanted forces, the application of wanted forces, and the application of desired motions" (Goertz, 1963).

Goertz's design was improved upon by Bergland in 1946 to handle the radioactive materials required by the Manhattan Project, and a major advance took place in 1949, when force feedback was added to the manipulator. Now the forces encountered by the slave arm were relayed to the operator by backdriving the master arm. This permitted the operator to "feel" the collisions of the slave arm with obstacles, providing better control. Another major advance was made when the mechanical linkages of the manipulators were replaced by electrical connections. This was accomplished by using variable resistance devices called potentiometers, or "pots," to measure the motion of the master or slave joints and transmit these electrical signals to servo motors used to drive the manipulator joints.

The next major advance was equipping these manipulators with a communications link, or "telephone." These devices, which are called "teleoperators," eventually permitted the modern use of remote manipulators in outer space (Heer, 1973). The teleoperator concept was extended a great deal by the Jet Propulsion Laboratory (JPL) researchers for the National Aeronautics and Space Administration (NASA) in their search for a general-purpose, dexterous machine for their space experiments. The machines had to be operable over extreme distances and provide for precise control. NASA and the Department of Energy used and continued the development of teleoperators into very sophisticated devices. Various remote manipulator experiments are shown in Figure 2–2. NASA now officially describes a teleoperator as a general-purpose, dexterous, cybernetic machine. A human controller commands the device with the aid of controls and displays. The device is located in a remote environment and has actuators to respond to human commands and sensors to feed information back to the human. The sensors might be television (TV) or force, auditory, or tactile sensors. The barrier between the human and the device can be a concrete wall, such as would occur in the handling of radioactive materials, or extreme distance, as in space applications.

The first machines on the Moon and Mars were teleoperator devices, which helped pave the way for Neil Armstrong and other astronauts. However, there was a problem caused by the long communication delay times between the human operator and the teleoperator. Even for the Lunar Lander, the delay time was too long to permit some operations that might otherwise have been performed. A delay time even as short as 1.3 seconds proved frustrating and difficult for the human operators. Therefore, the development of computer augmentation was necessitated. Local "autonomous reactions," such as those enabling the rover to stop before it fell into a hole, had to be done automatically. This evoked the addition of a computer to the remote device and sensors to permit it to develop "local reflexes," as shown in Figure 2–3. This was a major motivation for studies in intelligent machines.

To study this problem in a bit more detail, let's consider a control problem

Figure 2–2. Remote manipulators in experiments in various control modes. Top left: a slave manipulator equipped with stereo cameras. Bottom left: A mobile manipulator equipped with grippers and camera systems. Top right: The control boards showing the monitors for camera viewing, and the joysticks and switches for manipulator control. Bottom left: A side view of the mobile manipulator. (Courtesy of JPL/NASA.)

Remote Manipulators | **15**

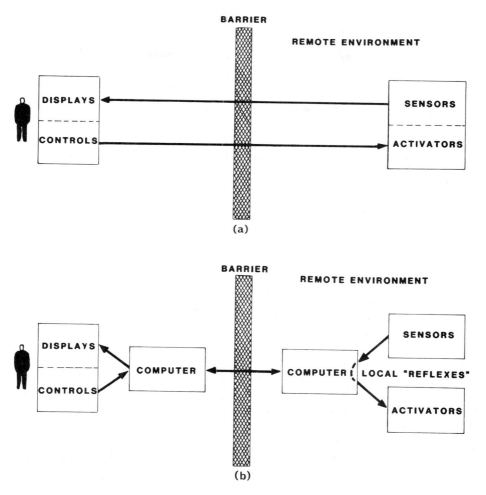

Figure 2–3. The basic concept of a master-slave manipulator. (a) The human, through the use of controls and displays, controls the activators and monitors the sensors in the remote environment. (b) The remote manipulator with local reflexes built into the manipulator. The use of some human control and some machine control is often required in difficult problems. (Courtesy of Microbot, Inc., Mountain View, California.)

encountered in the lunar experiments. The delay time in transmitting a signal from the Earth to the Moon was considered significant. Suppose the lunar rover was about to fall into a crater. If the transmission time for a STOP command took too long, the rover could plunge into the crater. To solve this problem, scientists at the Jet Propulsion Laboratory in Pasadena, California, started to design "local reflexes" or autonomous reactions into their teleoperators, such as that shown in Figure 2–4. This represents as important a step

16 | **History of Robots**

Figure 2–4. A robotic rover designed as a study tool to develop a rover to explore the surface of Mars or one of Jupiter's or Saturn's satellites. The rover is equipped with on-board computers that guide the robotic hand-eye system. The rover vision system under development consists of two television cameras mounted on masts above the chassis. The cameras allow the rover to interpret its surroundings, noting obstacles or objects of interest, and working in conjunction with a manipulator arm (lower left) to retrieve rocks. The rover, unlike other robots, will perform numerous tasks with a single command. (Courtesy of JPL/NASA.)

in the evolution of robots as it was in the evolution of animals. Many human reactions used in walking are controlled by local reflex control signals coming more from the spinal cord nerves than from the brain. Much of today's work in intelligent robots is in attempting to build local reflexes into our machines. Further advances, such as the addition of sensors and voice control, make today's remote manipulators quite sophisticated, such as those used at the Oak Ridge National Laboratory, shown in Figure 2–5. Other advances include using a mobile base, two arms, force feedback, audio feedback, stereo vision, computer control, and voice commands.

    The early robot manipulators were very simple. A human had to control their motions. The addition of a computer control to these manipulators extended their capabilities. The computer became the "brains" of the manipulator "body." Eventually,

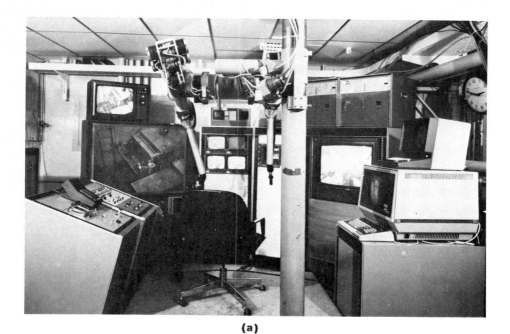

(a)

(b)

Figure 2–5. Modern remote manipulator developed at the Oak Ridge National Laboratory for material reprocessing. (a) Master manipulator and control room, which contains a lightweight pair of master arms, large screen and stereo viewing, voice control, and control computers. (b) Slave manipulator located in a remote hot cell. The bridge, carriage, and hoist provide 3 degrees of mobility. Each of the two manipulator arms has 6 degrees of freedom. The stereo mounted cameras have pan, tilt, zoom, and focus under voice control. The large number of degrees of freedom (19) makes the system difficult for a human to operate. The time required to complete a task can be 100 times greater than that required by humans. (Courtesy of Oak Ridge National Laboratory, Oak Ridge, Tennessee.)

computer and manipulator technology developed together to achieve the precise movements needed to perform many different kinds of work. Today's remote manipulators are still useful because they bridge the gap between operations that require some human control and operations that may be fully automated. Such tasks as the remote maintenance of equipment are being studied because these tasks are still too difficult for totally automated procedures. Some of the operational tasks are even too difficult for a single human and require complex control stations. Currently, robotics and remote handling in hostile environments are major research areas (*Proc. of Robotics and Remote Handling in Hostile Environments,* 1984).

2.5 Key Events in the Recent History of Robotics

In the 1940s, the war spurred the greatest government-industry-university cooperation ever achieved, and the results were astounding. This period was one of the most technologically productive eras in our history. Inventions that came from this era include modern communications, radar, sonar, great advances in automobile, aircraft, and ship manufacturing, the computer, and the atomic bomb. Computers as we know them today evolved rapidly.

Between 1940 and 1942, an automatic sequence controller was built at Harvard University. From 1943 to 1948, ENIAC, the first electronic computer, was built at the University of Pennsylvania. It was hardly the neat, compact device we know today, but a collection of electronic hardware that filled an entire room, which made it rather impractical for many applications. The next major invention that followed solved this problem.

At about the same time the manipulator was invented, a major electronic invention was made. In 1948, the transistor was invented by Bardeen, Bratton, and Shockley at Bell Telephone Laboratories. In the same year, EDSAC, the first stored program computer, was developed at Cambridge University.

Since computing capability could now be built into machines, the combination of the intelligence capabilities of the computer and the mechanical capabilities of the machines intrigued some of our greatest scientists. Claude Shannon was such a scientist. In 1952, he developed a robot mouse that could "learn" and run a maze. Shannon and his creation are shown in Figure 2–6.

At the same time that IBM ushered in the beginning of the computer age with its IBM 701 computer in 1952, an electromechanical feedback device called the "servo" was patented by George Devol. Devol's patents were to become the technical basis for the formation of Unimation, Inc., the first major robot manufacturer. In 1956, numerically controlled machine tools were offered by Cincinnati Milacron. These machines were programmed off-line, and a punched paper tape contained the commands for the machine. When the tape was read back into the machine tool, the programmed actions were carried out. In 1959, Planet Corporation offered the first commercial industrial robot (Ayres and Miller, 1983, p. 21). It was a pick-and-place device controlled by limit switches and cams. In 1961, Unimation introduced the first servo-controlled industrial

Key Events in the Recent History of Robotics | 19

Figure 2–6. Claude Shannon is shown here with his robot mouse, which could navigate a maze of 25 squares in 15 seconds. The robot "learned" its way through the maze after a 2-minute trial-and-error run. (Courtesy of the Massachusetts Institute of Technology Museum, Cambridge, Massachusetts.)

robot, similar in design to the robot shown in Figure 2–7. In the same year, INTEL Corporation was formed by Gordon Moore and Robert Noyce, which was to develop and market the first microprocessor. During this same time, work in plastics and electronics led to the previously mentioned prostheses (Ayres and Miller, 1983, p. 16). At the Massachusetts Institute of Technology, H. A. Ernst (1961), as part of his doctoral studies, connected a teleoperator slave arm equipped with touch sensors to a computer at the Lincoln Laboratory. This early connection of a computer and manipulator helped pave the way for future industrial robots.

In 1963, the American Machine and Foundry Company (AMF) introduced the VERSATRAN commercial robot. Starting in this same year, various arm designs for manipulators were developed, such as the Roehampton arm and the Edinburgh arm.

Meanwhile, other countries (Japan, particularly) began to see the potential of industrial robots. As early as 1968, the Japanese company, Kawasaki Heavy Industries, negotiated a license from Unimation for its robots. Japan's enthusiasm for the robot has since been astronomical.

One of the more unusual developments in robots occurred in 1969, when an experimental walking truck was developed by General Electric Company for the U.S. Army. This device is shown in Figure 2–8. Its control proved to be a very difficult problem even for a human and encouraged more investigation into automatic control. The large number of degrees of freedom required in the four-legged devices was a

20 | History of Robots

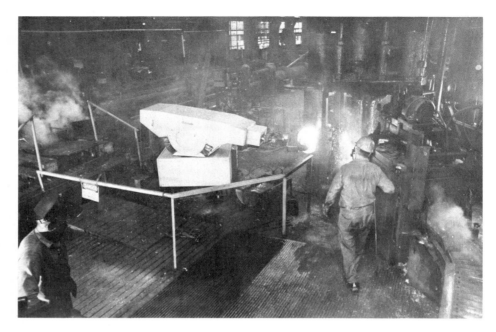

Figure 2–7. A Unimate robot similar to the first servo-controlled industrial robot, working in a hazardous industrial foundry application. This robot is unloading hot ingots from a furnace. (Courtesy of Joseph F. Engelberger.)

fundamental problem in control. In the same year, the Boston arm was developed, and the following year, the Stanford arm was developed, which was equipped with a camera and computer controller, and some of the most serious work in robotics began as these arms were used as robot manipulators. One experiment with the Stanford arm consisted of automatically stacking blocks according to various strategies. This was very sophisticated work for an automated robot at this time.

In 1970, the first national meeting for roboticists, the National Symposium on Industrial Robots, was held in the United States. In 1971, the Japan Industrial Robot Association was formed to foster the use of robots. In 1974, Cincinnati Milacron introduced the first computer-controlled industrial robot. Called "The Tomorrow Tool," or T3, it could lift over 100 pounds as well as track moving objects on an assembly line.

In 1975, the Robot Institute of America was formed from manufacturers and users of industrial robots. The British Robot Association was formed in 1977. In 1981, Robotics International of the Society of Manufacturing Engineers was formed as an individual member organization for those interested in robots, offering various certification and educational programs.

In barely 20 years, industrial robot installations in the United States went from zero to over 6000. Like the confluence of rivers, mechanical, electrical, and industrial technology combined to produce the modern industrial robot. As work in intelligent

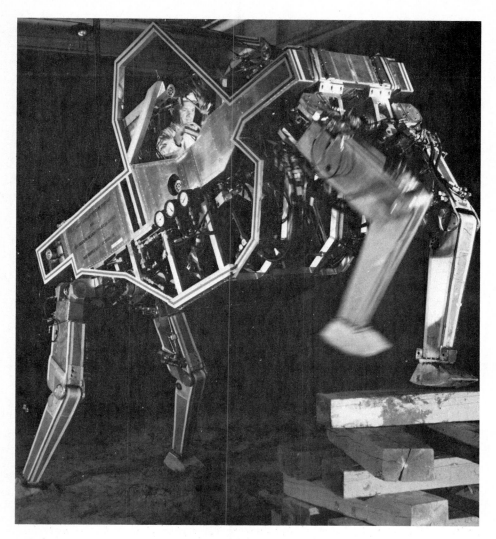

Figure 2–8. Research prototype of a four-legged quadruped machine, fabricated by General Electric Company engineers under a U.S. Army contract, was designed to spur development of equipment to improve the mobility and materials-handling capabilities of the foot soldier under the most severe conditions. By means of an advanced control system, the machine mimics and amplifies the linear movements of its operator. The right front leg of the unit is controlled by the operator's right arm, its left front leg by the operator's left arm, its right rear leg by the operator's right leg, and its left rear leg by the operator's left leg. The research prototype, 11 feet high and 3000 pounds in weight, was built by the GE Specialty Materials Handling Products Operation under a project sponsored jointly by the Advanced Research Projects Agency, Department of Defense, and the Department of the Army. The U.S. Army Tank-Automotive Command has acted as the contracting agency and has provided technical supervision of the project. The walking remote manipulator proved to be a difficult machine to control because of the large number of degrees of freedom required by the human operator. (Courtesy of General Electric Company, Bridgeport, Connecticut.)

robots makes these machines evermore versatile and capable, it is possible that robot installations could increase very rapidly in the coming years. Robotics has advanced from science fiction to reality in an amazingly short time. It is not hard to project that future robotics developments will exceed anything that has yet been anticipated, especially in the area of intelligent robots. Computers are becoming smaller, smarter, and cheaper every year. Thus, robots are also to become smaller, smarter, and cheaper as time goes on. New advances in electronics, computers, controls, and power systems will have significance for robot designers. There appears to be no limit to the mechanical applications of specialized robots. Perhaps the robot will rank with the clock someday in changing the way we perceive our world.

Chronology of Events in the History of Industrial Robots

Time	Technology
3000 B.C.	Egyptian water clocks and figures.
500 B.C.	Herodotus describes the wooden foot of Hegesistratus.
218 B.C.	Roman general Marcus Sergius has an iron replacement made for his severed hand.
1400s	Swiss and German android clocks developed.
1500–1700	**Scientific revolution**
1509	Götz von Berlichingen's iron hand is made with gearing for manipulating mechanical fingers and thumb.
1720	Bouchon and Falcon in Lyons, France, design looms for weaving patterns into silk.
1750–1850	**Industrial Revolution**
1770	Android automatons reach their peak.
1800	Metal lathe invented by Henry Maudslay.
1805	Joseph Marie Jacquard invents an automatic loom with punched card control.
1818	Milling machine invented by Eli Whitney.
1823	Construction of Charles Babbage's difference engine for calculating mathematical tables for navigation is begun.
1892	Motorized rotary crane with gripper for removing ingots invented by Steward Babbitt (U.S. Patent 484,870).
1930s	Spray-painting machines with recorded paths invented.
1940	**Government-industry-university cooperation**
1940–1942	Automatic sequence controller developed at Harvard.
1943–1948	ENIAC, the first electronic computer, developed at the University of Pennsylvania.
1944	Master-slave manipulator invented by Goertz (U.S. Patent 2,695,715, 1954).
1946	Electromechanical feedback manipulator invented by George Devol (U.S. Patent 2,590,091, 1952).
1948	Transistor invented by Bardeen, Bratton, and Shockley at Bell Laboratories.
1949	EDSAC, the first stored program computer, developed at Cambridge University.
	Force feedback added to remote manipulators.
1952	IBM's first commercial computer, the IBM 701, is marketed.

Chronology of Events in the History of Industrial Robots (continued)

Time	Technology
1956	Numerical control machine tool introduced by Cincinnati Milacron.
1959	First commercial robot is introduced by Planet, a pick-and-place device controlled by limit switches and cams.
1961	Unimation introduces the first servo-controlled industrial robot (Devol, U.S. Patent 2,988, 237).
	INTEL is formed by Gordon Moore and Robert Noyce.
	Collins prosthetic hand developed.
	Ernst arm, a teleoperator slave arm equipped with touch sensors, is connected to a computer at MIT's Lincoln Laboratory.
1963	AMF's VERSATRAN commercial robot introduced. American Machine and Foundry Versatile Transfer developed (Prab).
1963–1976	Roehampton arm developed.
	Edinburgh arm developed.
1968	Kawasaki Heavy Industries negotiates license from Unimation.
1969	Experimental walking truck is developed by General Electric for the U.S. Army.
	Boston arm developed.
1970	Stanford arm, with arm, camera, and computer connected that could stack colored blocks.
	First National Symposium on Industrial Robots.
1971	Japan Industrial Robot Association formed.
1974	Cincinnati Milacron introduces the T3, the first computer-controlled industrial robot, which could track objects on a moving conveyor.
1974	ASEA introduced electric drive industrial robot.
1975	Robot Institute of America formed.
1976	Viking II lands on Mars.
1977	British Robot Association formed.
1980	Fujitsu Fanuc Company of Japan develops automated factory.
	MAZAK flexible manufacturing factory is built in Florence, Kentucky.
1981	Robotics International/SME formed.
1982	First educational robots introduced by Microbot and Rhino.

Questions

1. What is a universal machine?

2. What was the main precursor of the modern industrial robot?

3. When was the first industrial robot offered commercially?

4. What is a teleoperator?

5. What are some of the problems roboticists face in robot control, and how do you think they might be overcome?

Robot Components and Operation

3.1 Basic Components

Industrial robots of today come in a variety of sizes, shapes, and capabilities. All have three basic components: a manipulator, a computer controller, and a power source, as shown in Figure 3–1. This illustration shows a modern industrial robot with a humanlike or anthropomorphic manipulator arm, a microcomputer controller, and a power source, which consists of both electric and hydraulic components. The manipulator arm moves tools or material to do the work. The arm shown has a load capacity of 225 pounds and can reach 13 feet. The controller provides the sequence of control signals to the manipulator and provides for interface to sensors, such as switches and cameras, to determine the motion the robot performs. The power source, which is usually electric, hydraulic, or pneumatic, provides the energy for the robot motions. In space, solar or nuclear power sources may be required. As yet, no hydrocarbon-powered (which we humans are) robots have been suggested, except in science fiction.

To use, program, or design an industrial robot, we must understand how they move and manipulate objects.

Manipulators

Since we live in a three-dimensional world, the general robot must be able to reach a point in space. Such points are described by coordinates. The robot may move forward and backward, to the left and right, and up and down. This may be accomplished in several ways. The simplest conceptual form moves independently, although concurrently, in three mutually perpendicular directions in a manner named after the famous French mathematician, René Descartes, who first developed this form of defining a point in space. This is the Cartesian, rectilinear, rectangular coordinate, or (x,y,z)-geometry robot shown in Figure 3–2a. The first coordinate, x, might represent left and right

Basic Components | 25

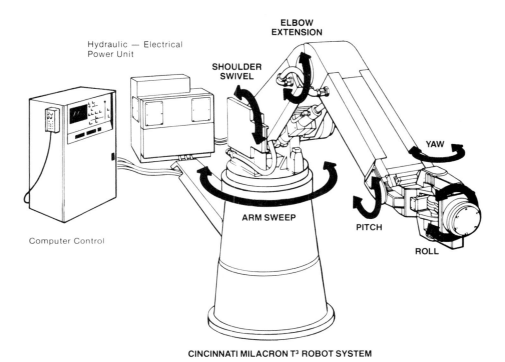

Figure 3–1. An industrial robot consists of three major components: a manipulator arm, a controller, and a power source. The 6 degrees of freedom—arm sweep, shoulder swivel, elbow extension, wrist pitch, yaw, and roll—are sufficient to place a tool in any position and orientation in the robot's work space. (Courtesy of Cincinnati Milacron.)

motion; the second, y, may describe forward and backward motion; the third, z, generally is used to depict up-and-down motion. The advantage of this design form is that motions in one direction can be made independently of the other two. Also, equal increments of motion may be achieved in all axes by using identical actuators (electric, hydraulic, or pneumatic motors). In general, the longer the arm, the less its stiffness. The locus or path of points that can be reached by a robot is called its work space or volume. The work volume of a Cartesian robot is a cube, so that any work performed by the robot must only involve motions inside this space. The work envelope of a robot is the outline of the work volume region. For a cubic work volume, the work envelope when viewed from either the top or side appears as a square. An actual Cartesian robot is shown in Figure 3–3. It also uses linear motions to move from one point to another.

Rotation of the manipulator about some axis gives the robot a simple method for moving around in a plane. Robots that have one rotational capability or degree of freedom and two translational (linear) degrees of freedom are called cylindrical coordinate robots. A degree of freedom is simply a variable motion. The first coordinate

26 | Robot Components and Operation

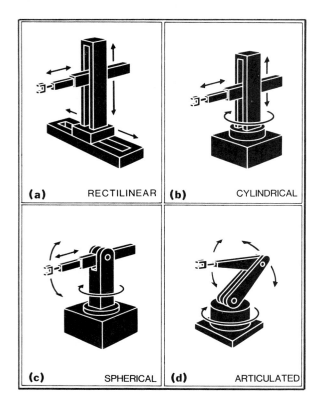

Figure 3–2. The four basic types of robots are described by their axes of motion. (a) A rectangular or Cartesian robot manipulator has three linear axes of motion and a cubic-shaped work volume. (b) A cylindrical or post-type robot has two linear motions and one rotary motion. The work volume of this robot has a cylindrical shape with the central core removed to accommodate the robot base and perhaps a pie-shaped section removed to provide for the backward extension of the arm. (c) A spherical or polar coordinate robot has one linear motion and two rotary motions. The work volume is shaped like a section of a sphere with upper and lower limits imposed by the angular rotations of the arm. A central core of the work volume is omitted to accommodate the robot base. A pie-shaped section may also be omitted to accommodate the rearward motion of the arm or to provide a safe operating position for the operator. (d) A jointed or anthropomorphic (humanlike) robot that uses three rotary motions. The work volume shape is spherical when viewed from the side and cylindrical when viewed from the top, with scallops on the inside limits of motion. (Courtesy of Cincinnati Milacron.)

describes the angle of base rotation, perhaps about the up-down, or z, axis. The second coordinate may correspond to a radial, or in-out, motion at whatever angle the robot is positioned. The final coordinate again corresponds to the up-down, or z position. The cylindrical coordinate robot shown in Figure 3–2b can reach any point in a cylindrical volume of space, although a central portion of the space must be devoted to the robot,

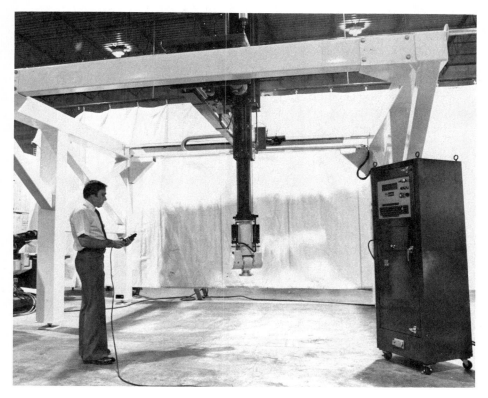

Figure 3-3 A Cartesian robot developed by GCA Corporation. (Courtesy of GCA, St. Paul, Minnesota.)

and limits to the full rotation may also be imposed. Its rotational ability gives it the advantage of moving rapidly to a point in the z plane of rotation. An actual cylindrical coordinate robot is shown in Figure 3–4. The resolution of a cylindrical robot is not usually equal in its three axes of motion. The resolution of the base rotation is expressed in terms of an angular measurement, and the linear axes' resolution is expressed in terms of linear increments.

The spherical coordinate robot, shown in Figure 3–2c, reaches any point in space through one linear and two angular motions. The first motion corresponds to a base rotation about a vertical axis. The second motion corresponds to an elbow rotation. The third motion corresponds to a radial, or in-out, translation. The two rotations can point the robot in any direction and permit the third motion to go directly to a specified point. The points that can be reached by the spherical coordinate robot include the volume of a globe or sphere. An actual spherical coordinate robot is shown in Figure 3–5.

The anthropomorphic or jointed-arm robot shown in Figure 3–2d uses three rotations to get to any point in space. This design is similar to the human arm, which has

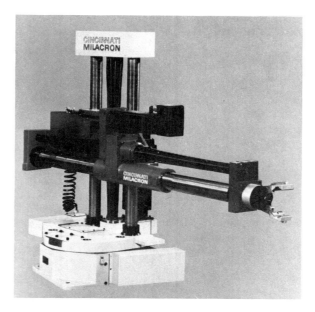

Figure 3–4 A cylindrical coordinate robot developed by Cincinnati Milacron, Model T3 363. (Courtesy of Cincinnati Milacron.)

two links—the shoulder and the elbow—and positions the wrist by rotating the base about the z axis, then rotation of the shoulder, and finally rotation of the elbow. In the jointed-arm robot, the first rotation is about the base and is a rotation about the z axis. The second shoulder rotation is a rotation about a horizontal axis. The final motion is a rotation of the elbow, which may be a rotation about a horizontal axis, but the axis may be at any position in space determined by the base and shoulder rotations. For the jointed-arm robot, the work envelope when looked at from the top of the robot is circular. When looked at from the side, the envelope has a circular outer surface; however, the inner surface has scallops due to the limits of the joints. An actual anthropomorphic robot is shown in Figure 3–6. The jointed-arm design can move at high speeds in various directions and has a greater variety of angles of approach to a given point.

The Cartesian robot has a cube-shaped work volume. The cylindrical robot has a cylindrical work volume. The spherical robot has a spherically shaped work volume. The anthropomorphic robot has a somewhat spherically shaped work volume with scallops due to the joint constraints, as shown in Figure 3–7. Each practical robot has fixed limits of motion so that the actual work volume is usually less than the theoretical volume. For example, some space must be left out of the work volume for the robot manipulator, since no two solid objects can occupy the same space simultaneously. However, each design has found wide application.

It is interesting to compare the jointed-arm robot to the human arm. First, we have a

Basic Components | 29

Figure 3–5. A Unimate spherical coordinate robot. (Courtesy of Joseph F. Engleberger.)

mobile base through the use of our legs, which may be used to position our body at any accessible point in space. Next, we have 3 degrees of rotation more at our waist, which might be used to orient the shoulder in any direction. These extra or redundant degrees of freedom give dexterity to the mobile human over the single base rotation of the robot. At the shoulder, which we may compare to the industrial robot shoulder, we have a very unique ball-and-socket joint. Instead of a single rotational degree of freedom, the human shoulder can rotate in three different ways. It can rotate up and down about a horizontal axis, which we may call pitch. It can rotate forward and backward about a vertical axis,

30 | **Robot Components and Operation**

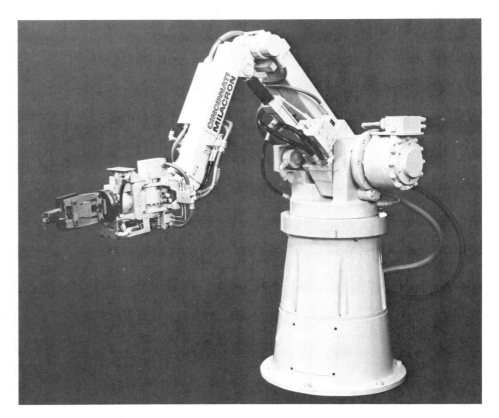

Figure 3–6. An anthropomorphic robot. (Courtesy of Cincinnati Milacron.)

which we may call yaw. Finally, it can roll about an axis along the straightened arm. Now, let's consider the elbow rotation with the human in a sitting position. The human elbow can rotate about a horizontal axis, a vertical axis, and about the axis of the forearm, achieving pitch, yaw, and roll actions. These extra degrees of freedom in the human arm are redundant, since only three are needed to position the wrist at any point in our work space. They give humans many different ways to position the wrist in space, aid in avoiding collisions, and improve the ability to reach into constrained spaces. You may want to try the following experiment. Close your fist, and keep the wrist joint rigid. Pick any point in space, say, an object on a table in front of you within your reach. Now, explore the many ways you can move your hand toward that point. An infinite variety of pathways can be used. This fascinating versatility of the human has not yet been built into our industrial robots. Perhaps the most basic reason is cost. Each degree of freedom requires an actuator, power source, and control mechanism. Another reason is simply that we do not yet have the mathematical sophistication to solve the equations of motion for such a complicated mechanism as the

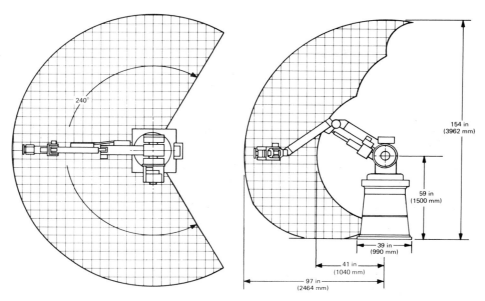

Figure 3–7. The work volume of a jointed-arm robot. (Courtesy of Cincinnati Milacron.)

human arm, nor do we have the computer power to make the control calculations at a nearly human speed. Furthermore, no one has yet designed a remotely actuated ball joint, such as that in the human shoulder. Fundamental and applied research are needed in the basic mathematics required for the control, in high-speed computation, and in electromechanical design before we can fully understand or duplicate the human arm. In the meantime, it is a tribute to robot designers and engineers that they have developed designs and practical industrial robots that perform so well.

The arm may be driven or actuated directly with the actuator located at the joint, or remotely with the actuator located at the base and a transmission used to transmit power to the joint. For an anthropomorphic design, 3 rotary degrees of freedom are required. The base can often be directly driven, since its actuator can be located in the base itself, and therefore not add to the weight and inertia of the manipulator. The shoulder actuator may also be located in or near the base. The elbow actuator may require careful design to avoid adding significant inertia to the base. Piston and cylinder, ball screws, gear drives, or other remote drives are often used to minimize the weight and inertia of the arm.

A new form of robot has recently been developed. Its design is somewhat like the links in our spine or of the vertebrae in a snake. This robot is called the active cord mechanism (ACT) by its developer, Professor Y. Umetani, of Tokyo University. The "spine" design provides a very flexible motion characteristic that would even permit it to move through a pipelike space.

Mobility may be considered a separate attribute of the robot manipulator. Mobile robots perform useful tasks by moving themselves and perhaps a payload through various programmed motions. This class of robots may include wheeled, tracked, or leg-type robots.

Wrists

Thus far, we have described the robot arm's 3 degrees of freedom that permit it to position a tool at any point in three-dimensional space within its work volume. To perform useful work, it must also be able to position a tool in any possible orientation. This requires a wrist, which in general requires 3 degrees more of rotational freedom for tool orientation.

Wrists may be designed with various degrees of freedom and in various configurations. Stackhouse (1979) developed an interesting classification for robot wrists based upon the type and sequence of rotations used to obtain the three rotary motions. Three rotations are required to orient a tool or part in any desired manner. To appreciate this, suppose you are working with a regular screwdriver with a blade that is thick in one dimension and thin in the other. To orient the screwdriver requires a pitch rotation up and down about a horizontal axis to provide the first angle of orientation. Another rotation, yaw, around another horizontal axis is required for the second. Finally, to align the blade of the screwdriver requires a final rotation, a roll, which is a rotation about the axis of the screwdriver. These three angular rotations would also orient a tool, such as a gripper, a wrench, or a welding gun.

In Stackhouse's classification, a distinction is made between a bending rotation and a roll rotation. In the screwdriver example, the pitch and yaw rotations would be classified as bends, which are rotations about an axis perpendicular to the longitudinal axis of the link or tool. The final rotation is defined as a roll, which is a rotation about the link or tool axis. The difference in these rotations is that a bend motion is restricted from full 360-degree rotation by its link, but a roll rotation could theoretically rotate all the way around. We may also define the rotations from the workpiece or tool action. A pitch is a rotation about a horizontal axis. In an airplane, pitch causes the nose of the plane to move down or up. Yaw is a rotation about a vertical axis. In the airplane, a yaw motion moves the nose of the plane to the left or right. Finally, a roll is a rotation about the axis of the link. In the airplane, a roll motion turns the plane about its own axis. With these definitions in mind, let's consider various wrist designs that have 1, 2, or 3 degrees of freedom.

A variety of wrist actions are shown in Figure 3–8. Wrists with 1 degree of freedom could involve either a bend or a roll. However, we will see that on almost all industrial robots, the final rotation is a roll. Wrists with 2 degrees of freedom are more practical, because some robots, such as those used for spray painting, may need only 2 degrees of freedom to point a spray device in a given direction. These present two possibilities. Let's call group 1 wrists those that have two rolls. Two-roll wrists would have two separate links, each link capable of rotating about its longitudinal link axis. Consider the following two designs. In one design, which we will call type A, the axes of the two links intersect at a point that is offset from the physical point of contact. In type B, the intersection and the point of contact are coincidental.

Group	Wrist Axes	Wrist Orientations of Payload
1	Roll-Roll	Pitch & Yaw or Pitch & Roll
2	Bend-Roll	Pitch & Roll
3	Bend-Bend-Roll	Pitch, Yaw, & Roll
4	Bend-Roll-Roll	Pitch, Yaw, & Roll
5	Roll-Bend-Roll	Pitch, Yaw, & Roll
6	Roll-Roll-Roll	Pitch, Yaw, & Roll

(Note: By indexing the wrists in each group by 90° about the longitudinal axis of the arm, the pitch and yaw axes of orientation will become yaw and pitch axes, respectively.)

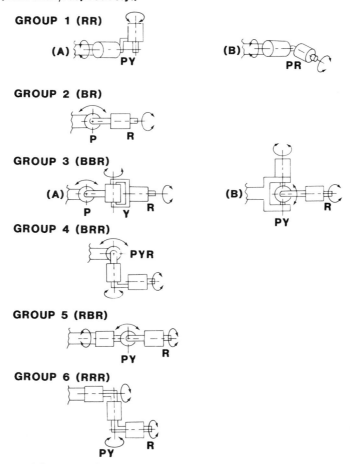

SCHEMATICS OF WRIST CONFIGURATIONS

Figure 3–8. A classification of wrist designs by the sequence, from the arm, of the rotations used in the wrist. Note that each wrist ends with a roll action. This is typical of manipulative wrists. Other wrist designs may be used for special applications, such as camera or other sensor mounts. (Courtesy of Cincinnati Milacron.)

Group 2 wrists are the other type of two-roll wrists, but in these designs, a bend is followed by a roll. The Microbot uses this type of design.

For wrists with 3 degrees of freedom, there are four possible design groups, each of which ends with a roll action. Group 3 wrists are of the bend-bend-roll variety. These implement pitch, yaw, and roll actions as seen from the manipulator arm. Type A has separate axis locations for the pitch and yaw motions. Type B has intersecting axes for the pitch and yaw actions.

The next wrist design, group 4, has a bend followed by two roll actions. Group 5 designs have a roll-bend-roll action. Group 6 wrists have three roll actions. Two design variations are possible within this group. Type A covers those in which the roll axes are not coaxial, and type B covers those in which the roll axes are coaxial. An example of this type of wrist is the Cincinnati Milacron design of its patented, three-roll wrist, used on its T3 746 robots.

The wrist may be driven directly with the actuator located at the joint or remotely with the drive power transferred from the base with chain drives, rigid links, or other mechanisms. The directly driven design can generally supply greater strength but adds to the weight and inertia of the manipulator. The remotely driven wrist reduces the inertia of the manipulator but adds to the complexity of the design since a transmission is required.

We have now considered the basic robot manipulator design that implements 6 degrees of freedom—3 that enable it to position a tool at any point in three-dimensional space, and 3 orientation angles, which permit it to position a tool in any orientation. Interestingly, almost all industrial robots have a roll action as the last motion, so that such motions as those required to turn a screwdriver can be accommodated. This design is different from the human arm, which implements its main roll capabilities in the elbow and shoulder.

The human arm has several more than the 6 degrees of freedom found on today's industrial robots. The design and use of these redundant degrees of freedom is an exciting topic for further research.

Let's look at some examples of commercial wrist designs. The Cincinnati Milacron T3 wrist is shown in Figure 3–9. This wrist is a group 3 design with two bend motions followed by a roll action. This wrist is moved by hydraulic actuators located near the rotational joints. Its initial design permitted it to lift a 100-pound load located 10 inches from its faceplate. This design permits great lifting capacity; however, it is directly rather than remotely driven. That is, the actuators are located at the joint action locations. This adds weight to the arm and makes the wrist rather bulky since the wrist size also includes the actuators.

A Unimate 4000 wrist design is shown in Figure 3–10. This is a group 4 design with a bend, roll, and roll action. The wrist is capable of lifting a 175-pound load and providing 3500 inch-pounds of torque for its first bend action, 2800 inch-pounds of torque about the first roll axis, and 2300 inch pounds of torque about the final roll axis. This wrist is also remotely driven and contains an ingenious set of gear trains in the wrist.

A schematic of the Renault robot wrist is shown in Figure 3–11. This wrist is a group 5 design since it has a roll, bend, roll configuration. This wrist is remotely driven

Basic Components | 35

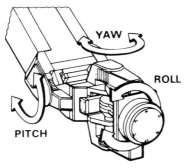

100 lb load at 10 inches
from the face plate

Figure 3–9. An example of a bend-bend-roll or group 3 wrist used on the Cincinnati Milacron T3 robot. The wrist is directly driven by the hydraulic actuators located on the wrist. (Courtesy of Cincinnati Milacron.)

by three actuators that transmit the drive force through a gear train, shown in the illustration, to achieve the pitch, yaw, and roll motions.

As a final example, let's consider the Cincinnati Milacron three-roll wrist shown in Figure 3–12a. This wrist fits into the group 6 category with three consecutive roll actions and with all three roll axes intersecting at one point. This patented, unique design was

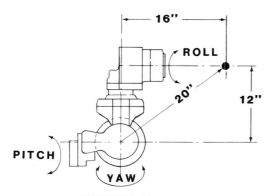

UNIMATE 4000 WRIST
175 lb load
Bend: 3500 in-lb
Yaw: 2800 in-lb
Roll: 2300 in-lb

Figure 3–10. An example of a bend-roll-roll or group 4 wrist design used on the Unimation/Westinghouse 4000 series robot. (Courtesy of Joseph F. Engleberger.)

RENAULT ROBOT WRIST

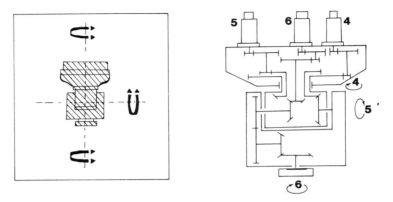

Figure 3-11. An example of a roll-bend-roll or group 5 wrist design used on a Renault robot. The wrist is remotely driven by actuators that activate a gear train located in the wrist.

reportedly conceived on a Saturday while Ted Stackhouse, an engineer at Cincinnati Milacron, was working in his basement. He looked up at the ductwork for his heating system and noted that the axes of the bent ductwork were coincident and realized that a robot wrist could be built using the same offset. The wrist also includes a uniquely simple gear train that could be remotely driven by three concentric torque tubes to provide the remote actuation for the three roll motions, as shown in Figure 3-12b. An outer torque tube drives the first roll, an middle torque drives the second roll, and an inner torque tube drives the final roll.

End Effectors

With a clear idea of the arm and wrist geometries that provide the general robot with the ability to reach any point in space with any orientation, it is now time to consider the "business end," or end effector, of the robot.

Robot end effectors come in great variety to provide versatility. Basically, end effectors can be divided into two types—grippers and process tooling. Grippers may be two- or more fingered devices designed to grasp an object or tool in a manner similar to the human hand and fingers. Process tooling may be any useful device, such as a spot-welding torch, a spray-painting gun, a vacuum suction cup, or a set of interchangeable tools.

Grippers may be designed as physical constraint or as friction devices. A physical constraint device might work like a spatula that slides under an object to enable one to lift it. A frictional device depends upon the frictional force between two materials to provide the gripping force. When you put your hand under an object to lift it, you are using the physical constraint principal. When you grasp an object with your fingers to lift it, you

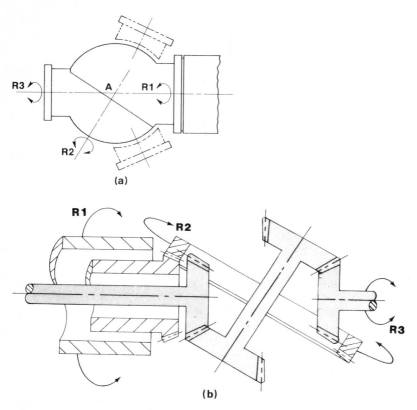

Figure 3–12. (a) An example of a roll-roll-roll or group 6 wrist used on the Cincinnati Milacron robot. The wrist actions are remotely driven by concentric torque tubes that power a gear train located in the wrist. (b) The wrist includes a unique, simple gear train. The gear train and torque tubes are shown here. (Courtesy of Cincinnati Milacron.)

are relying on the frictional force between the skin and object material to permit you to lift it.

When you lift an object using the constraint principle, it may appear that the only force you must overcome is the weight of the object. However, if you move the object, dynamic forces must also be overcome. Suppose you accelerate an object from a resting state at a rate equal to the acceleration of gravity. This is the rate at which an object falls in a vacuum. To get an idea of this acceleration, just drop a small steel ball. It will fall 16 feet in the first second. When you accelerate the object upward at this rate, you must not only overcome the static force, which is equal to its weight, but also overcome a dynamic force equal to its weight. The dynamic force equals the product of mass times acceleration. (The mass is equal to the weight divided by the acceleration of gravity. When this is multiplied by the acceleration of gravity, the result is simply the weight.) The total force to move the object in this example is equal to the sum of the static and dynamic forces, which is twice the weight of the object.

When you lift an object upward by gripping the object from its sides using frictional forces, only a fraction, given by the coefficient of friction, of the sideways or horizontal normal force is available to overcome the static and dynamic forces acting vertically. For example, if the coefficient of friction is 0.2, then a horizontal normal force equal to five times the weight is required to lift the object. (The required force equals the coefficient of friction divided by the vertical force.) Generally, the softer the material, the greater its coefficient of friction, and the more nearly equal the normal force will be to the vertical force. Again, moving the object would require a compensation for the dynamic force. To move an object upward at the acceleration of gravity always requires a normal force greater than its weight. If the coefficient of friction were 0.2, then a normal force five times its weight on each side would be required to accelerate the object. The total force required to lift and move the object would now be 10 times its weight. For example, if a 100-pound object were used, a normal force of 1000 pounds would be required. This same difference between constrained and frictional forces for gripping objects may partially explain why so many of our useful implements have handles on them.

Several examples of gripper designs are shown in Figure 3–13. The simplest design is shown in Figure 3–13a and is actuated by a linear actuator that pulls or pushes the two drive linkages that cause the gripping linkages to rotate and close on an object. The maximum clamping force, the size of the opening, and the moving speed of the gripper fingers depend directly on the location of the rotation centers. A four-linkage mechanism is shown in Figure 3–13b. The advantage of this design is that the fingers move in a parallel motion. This design is suitable for gripping box-shaped components. For handling soft objects, a design such as that shown in Figure 3–13c is appropriate. This frictional-type gripper has bellows on each side that are inflated to grasp the object. A dual gripper that can grasp either the inside or outside diameter of a part is shown in Figure 3–13d. A variation of this design for longer rectangular or cylindrical objects is shown in Figure 3–13e. Various commercial designs are shown in Figure 3–13f, including suction cups in different arrangements. The Skinner hand is shown in Figure 3–13g. This three-fingered hand, with 3 degrees of freedom for each finger, has 9 degrees of freedom and consequently much greater gripping flexibility than a two-fingered gripper. Finally, electromagnetic grippers are shown in Figure 3–13h. This great variety of gripper designs shows the flexibility that can be achieved in using robots for many applications. Also note that the multifingered hand is the best example yet developed of a universal gripper.

Process tooling refers to the general class of special end effectors that may be attached to the robot wrist. A spot-welding gun can be attached to the robot wrist to place a series of welds on flat or curved surfaces. Generally, a 3-degree-of-freedom wrist is required because of the dexterity required for maneuvering the gun. An arc-welding torch is another widely used end effector. The robot can position the welding torch for a single straight or curved run or use a weaving pattern for wider welds. Ladles are also used for applications in which the robot must scoop up and pour molten metal into a casting. Spray-painting guns are also commonly used by industrial robots. In some cases

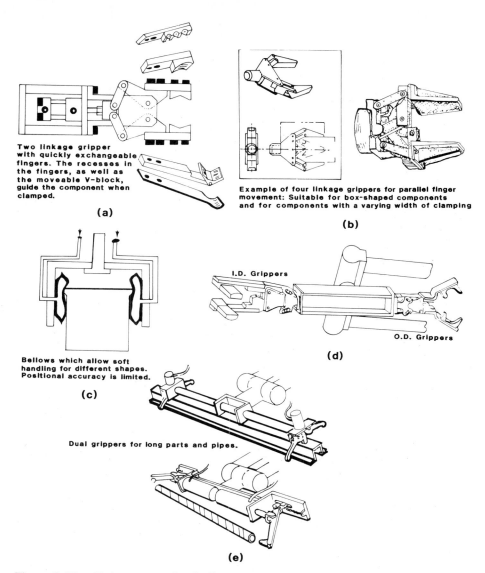

Figure 3–13. Various types of end effectors used on industrial robots. (a) A rotary or two-linkage gripper designed for interchangeable fingers. The linear motion of the drive mechanism causes the gripper linkages to rotate and open and close the gripper. The recesses in the fingers, as well as the movable V-block, guide the component when clamped. (b) An example of a four-linkage gripper design that provides for parallel finger movement. This gripper action is suitable for box-shaped components and for components requiring a variable width for clamping. (c) A bellows gripper that permits soft handling of objects of various shapes. The positional accuracy of such a design is limited. (d) A rotatable gripper designed to grasp either the inside or outside diameter of an object. (e) Elongated gripper designs suitable for grasping long objects from either the inside or outside diameters. (f) Various types of vacuum suction cups (Shinko, Japan.) (g) The Skinner hand, a three-fingered gripper that has 12 degrees of freedom. (h) Two forms of electromagnetic grippers. (Courtesy of Joseph F. Engelberger.)

40 | Robot Components and Operation

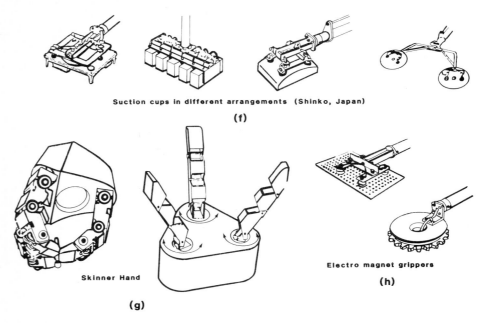

Figure 3–13 (continued)

only 2 degrees of freedom may be required of the robot wrist for spray painting. The robot can spray parts with compound curved surfaces. Grinders, routers, or sanders are also easily attached to a robot wrist. A large class of assembly tools, such as drills, screwdrivers, and wrenches, can be used by the robot. In some cases these tools are automatically interchangeable by the robot.

A variety of process tools is shown in Figure 3–14. These tools were developed by Unimation for the many applications of their industrial robots. A tool for ladling hot materials, such as molten metal, is shown in Figure 3–14a. This type of tool is used in casting applications. A spot-welding gun is shown in Figure 3–14b. The welding gun consists of electrodes that, when positioned by the robot, are energized to melt the materials to form a joint. Tools, such as impact wrenches, similar to those used to remove the nuts from automobile tire lugs may also be used for robotic applications. One such example is shown in Figure 3–14c. Other tools, such as drills, screwdrivers, and cutting tools, can also be attached to the robot wrist. A tool called a stud-welding head is shown in Figure 3–14d. Studs are fed through a tube and welded in place. An arc-welding torch is shown in Figure 3–14e. Arc welding is an important industrial application not only because it removes a human from a hazardous environment but also because it provides improved weld consistency and quality. The industrial robot can also manipulate a tool like the heating torch shown in Figure 3–14f. A tool for grinding is shown in Figure 3–14g. Grinding, edge routing, or sanding can be readily accomplished with an industrial robot with the appropriate tooling. Spray painting or adhesive

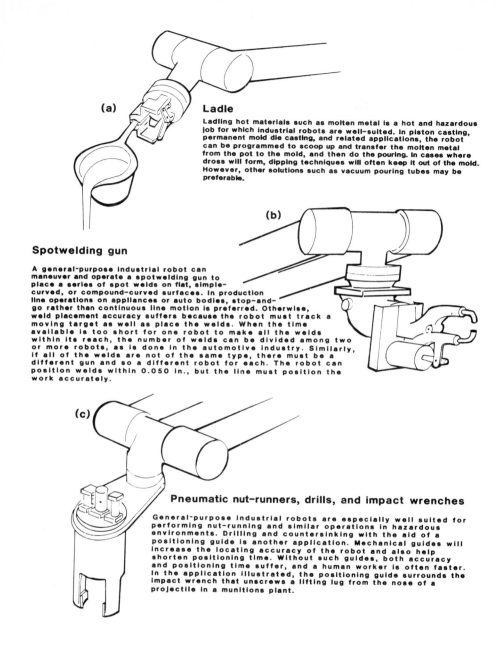

(a) Ladle

Ladling hot materials such as molten metal is a hot and hazardous job for which industrial robots are well-suited. In piston casting, permanent mold die casting, and related applications, the robot can be programmed to scoop up and transfer the molten metal from the pot to the mold, and then do the pouring. In cases where dross will form, dipping techniques will often keep it out of the mold. However, other solutions such as vacuum pouring tubes may be preferable.

Spotwelding gun

A general-purpose industrial robot can maneuver and operate a spotwelding gun to place a series of spot welds on flat, simple-curved, or compound-curved surfaces. In production line operations on appliances or auto bodies, stop-and-go rather than continuous line motion is preferred. Otherwise, weld placement accuracy suffers because the robot must track a moving target as well as place the welds. When the time available is too short for one robot to make all the welds within its reach, the number of welds can be divided among two or more robots, as is done in the automotive industry. Similarly, if all of the welds are not of the same type, there must be a different gun and so a different robot for each. The robot can position welds within 0.050 in., but the line must position the work accurately.

Pneumatic nut-runners, drills, and impact wrenches

General-purpose industrial robots are especially well suited for performing nut-running and similar operations in hazardous environments. Drilling and countersinking with the aid of a positioning guide is another application. Mechanical guides will increase the locating accuracy of the robot and also help shorten positioning time. Without such guides, both accuracy and positioning time suffer, and a human worker is often faster. In the application illustrated, the positioning guide surrounds the impact wrench that unscrews a lifting lug from the nose of a projectile in a munitions plant.

Figure 3–14. Various types of process tools. (a) Ladle for pouring hot materials, such as molten metal, into a mold. (b) Spot-welding gun used to place a series of welds to join two materials. (c) Pneumatic impact wrench that was used for unscrewing a lifting lug from the nose of a projectile in a munitions plant. (d) Stud-welding head. (e) Arc-welding torch. (f) Heating torch used to bake foundry molds. (g) Grinder used for removing rough edges from castings. (h) Spray-painting gun. (i) Changeable tools can be used with the industrial robot. In the application shown, spot- and arc-welding guns may to changed using the holding device. (Courtesy of Joseph F. Engelberger.)

Stud-welding head

Equipping an industrial robot with a stud-welding head is also practical. Studs are fed to the head from a tubular feeder suspended from overhead. One caution concerns accuracy with which welded studs can be located. An industrial robot can position a stud within 0.050 in., but on-the-line work positioning must be exact. The weight of the head is rarely a significant limitation. Stud-welding heads are well within the 100-lb capacity of standard robots.

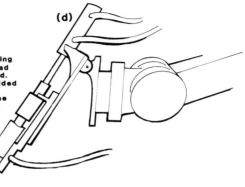

(d)

(e)

Inert gas arc welding torch

Arc welding with a robot-held torch is another application in which an industrial robot can take over from a man. The welds can be single or multiple-pass. The most effective use is for running simple-curved and compound-curved joints, as well as running multiple short welds at different angles and on various planes. Maximum workpiece size is limited by the robot's reach, unless the robot is mounted on rails. Where the angle at which the gun is held must change continuously or intermittently, the industrial robot is a good solution. But long welds on large, flat plates or sheets are best handled by a welding machine designed for that purpose. In addition to welding for fabrication purposes, wear-resistant surfaces and edges can be prepared by laying down a weld bead of tough, durable alloy. And the robot will handle a flame cutting torch with equal facility.

Heating torch

The industrial robot can also manipulate a heating torch to bake out foundry molds by playing the torch over the surface, letting the flame linger where more heat input is needed. Fuel is saved because heat is applied directly, and the bakeout is faster than it would be if the molds were conveyed through a gas-fired oven.

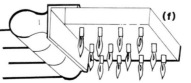

(f)

Figure 3–14 (continued)

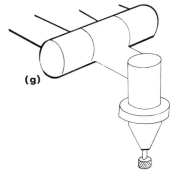

Routers, sanders and grinders

A routing head, grinder, belt sander, or disc sander can be mounted readily on the wrist of an industrial robot. Thus equipped, the robot can rout workpiece edges, remove flash from plastic parts, and do rough snagging of castings. For finer work, in which a specific path must be followed, the tool must be guided by a template. The template is a substitute for the visual-and sometimes tactile- control that a human worker would exercise. In such a case, the overall accuracy achieved depends upon how accurately the workpiece is positioned relative to the template. Usually, the part is automatically delivered to a holding fixture on which the template is mounted.

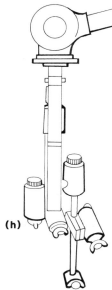

Spray gun

Ability of the industrial robot to do multipass spraying with controlled velocity fits it for automated application of primers, paints, and ceramic or glass frits, as well as application of masking agents used before plating. For short or medium-length production runs, the industrial robot would often be a better choice than a special-purpose setup requiring a lengthy changeover procedure for each different part. Also, the robot can spray parts with compound curvatures and multiple surfaces. The initial investment in an industrial robot is higher than for most conventional automatic spraying systems. When the cost of frequent changeovers is considered, the initial investment assumes less importance. Industrial robots can be furnished to meet intrinsically safe standards for installation in solvent-laden, explosive atmospheres.

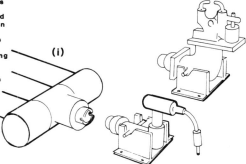

Tool changing

A single industrial robot can also handle several tools sequentially, with an automatic tool-changing operation programmed into the robot's memory. The tools can be of different types or sizes, permitting multiple operations on the same workpiece. To remove a tool, the robot lowers the tool into a cradle that retains the snap-in tool as the robot pulls its wrist away. The process is reversed to pick up another tool.

Figure 3–14 (continued)

placement can be applied by a tool such as that shown in Figure 3–14h. The device shown has two containers that may be used to paint in two different colors. Finally, an example of tool changing is shown in Figure 3–14i. Obviously, the variety of process tools that can be connected to the industrial robot is limited only by the applications required.

Compliant End Effectors. A special end effector that is neither a gripper nor a process tool but rather a device that fits between the robot wrist and end effector for special assembly applications is the remote center compliant (RCC) device. This device was developed at the Charles Stark Draper Laboratory of Cambridge, Massachusetts and is now commercially available. In effect, the RCC is a springy wrist attachment that permits parts to be mated together. The design permits the frictional forces encountered when putting a peg in a hole to rotate or translate the peg to fit the hole. The two errors that can be encountered in parts mating are a translational error when the part is not centered over the hole and a rotational error when the part is not aligned with the hole. The RCC device consists of two linkage mechanisms, each of which reduces one of the error types. The two mechanisms and an actual RCC device are shown in Figure 3–15. The rectangular linkage mechanism causes a translation in response to frictional forces. The trapezoidal portion produces a rotation about a remote center point, which aligns the part. In the commercial version both mechanisms fit together to produce a rugged, compliant design. Other devices for assembly are also available. For example, engineers at the Kawasaki laboratories in Japan can put together complex parts, such as motors and gearboxes, using high-precision feedback, cleverly designed grippers, and compliant fixtures. Also, an active RCC device has been constructed in which the center may be changed.

Control Units

The arm, wrist, and end effector compose the main moving portion of a robot. However, a controller and power source are also required. A controller may be as simple as a sequencer with a series of adjustable mechanical stops or as complicated as a hierarchical array of computers. The modern controller must contain memory to permit storage of programs and data, drive elements for each of the robot's degrees of freedom, and interface elements to permit the response to external signals. The robot can be programmed to accurately follow a path from point to point or a continuous path.

The modern programmable controller used with an industrial robot is a special microprocessor equipped with easily interfaced circuits for connections to external devices, a special programming language, and an industrial-quality enclosure. Methods of control are more fully described in Sections 3.2, 3.3, and 3.4.

Power Units

A power unit is required to move the robot and payload through the desired motion. A hydraulic power source is generally used for lifting heavy weights or in a possibly

Problems in Small Parts Mating

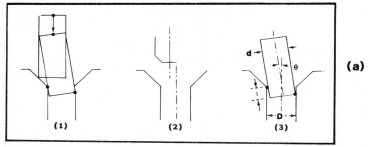

As part slides down chamfer (1) it acquires angular error if it is held at the top. Chamfer of peg falls within chamfer of hole (2). Only one chamfer on peg or hole is necessary. Defining terms for analysis of two point contact (3).

Remote Center Compliance (RCC) Device

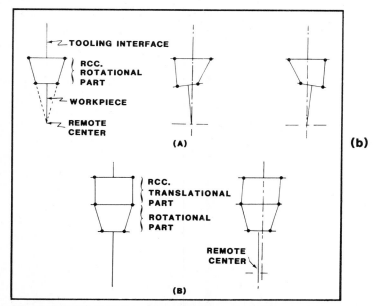

Sketches in (A) show two-dimensional representation of rotational part of RCC. Translational part allows workpiece to translate without rotating (B).

Figure 3–15. Problems in assembling small components can be overcome using the principle of remote center compliance. (a) Positioning errors result in an offset of the part from the hole. Orientation errors result in angular offset. (b) Design of a passive compliant device that uses only frictional forces to overcome both positional and orientational errors. The trapezoidal four-bar linkage causes the part to rotate about a remote center and align the part with respect to the hole. The rectangular four-bar linkage translates the part to overcome positional errors. (c) A schematic of a practical implementation that includes both compliant mechanisms in a compact arrangement. (d) A commercial remote center compliance device in a rugged compact design. (Courtesy of the Charles Stark Draper Laboratory, Inc., Cambridge, Massachusetts.)

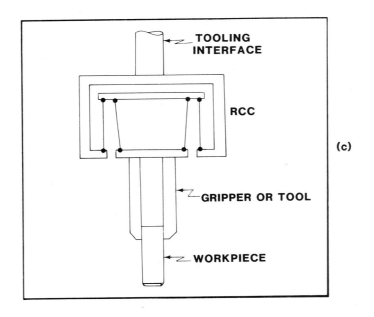

(c)

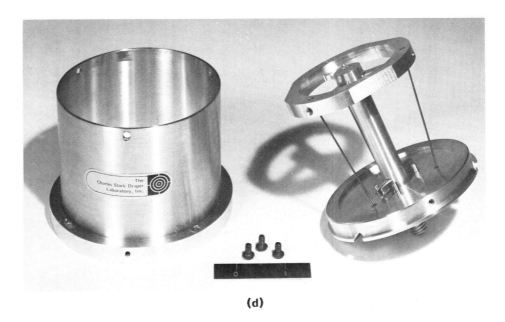

(d)

Figure 3–15 (continued)

explosive environment, such as in spray painting. An electric-drive robot may be used in a medium-weight application for which high accuracy is required. A pneumatic source has often been used for simple pick-and-place robots, especially in factory environments that have a readily available compressed air supply. A typical hydraulic power system is shown in Figure 3–16. In operation, the pump generates a flow at a given pressure. An accumulator unit is used to dampen any changes in the pressure due to loading or other factors, such as temperature changes. The fluid flows through a load, such as an actuator valve, and back through a cooler into a reservoir, where it is filtered and recycled. Hydraulic actuators can provide power to lift very heavy loads and provide smooth operation. Spray-painting robots, such as the DeVilbiss/Trallfa TR-4500 shown in Figure 3–17, typically use hydraulic power for these reasons and to avoid any explosive hazard that might be caused by an electric motor.

Many of the newer robots are using electric power in such applications as welding, in which there is no spark hazard. An example of an electrically powered robot is shown in Figure 3–18. Direct current motors may be designed to meet a wide range of power requirements and are relatively inexpensive and reliable.

Pneumatically powered robots have an industrial advantage because air power lines are usually as common in factories as are electrical lines in a home. Therefore, these robots can be installed without a separate power source by direct connection to the existing pneumatic plant power. Pneumatic action is especially popular for nonservo robots. An example of a pneumatically powered robot is shown in Figure 3–19.

3.2 Operation of Industrial Robots

We have just briefly considered the mechanical aspects of the industrial robot, including the manipulator, control unit, and power source. We will now consider the operation of the robot. In general, control systems may be divided into two types, commonly called open-loop or closed-loop systems. Both are used in industrial robots. An example of an open-loop system is a stepper motor in which the control signals directly position the motor without feedback. Two types of closed-loop systems are used. These two types are called nonservo and servo. Each uses a feedback signal.

In an open-loop system, the input is applied and the output behaves in accordance with the characteristics of the system. An example of such a system would be the heating of a house by a fireplace. The input to the system is the fuel put into the fireplace, and the output is the temperature of the house at some point in time. The output temperature is determined by the amount of fuel burned, the outside temperature, the wind velocity, the amount of insulation in the house, and other factors. In contrast, a closed-loop control for a house might be one that uses a thermostat with a desired temperature setting and a thermometer. If the house temperature is less than the desired temperature, the furnace is turned on until the temperature reaches the desired value. The cycle of operation is closed because the temperature of the house acts through the thermostat to operate the furnace that heats the house and produces the desired temperature.

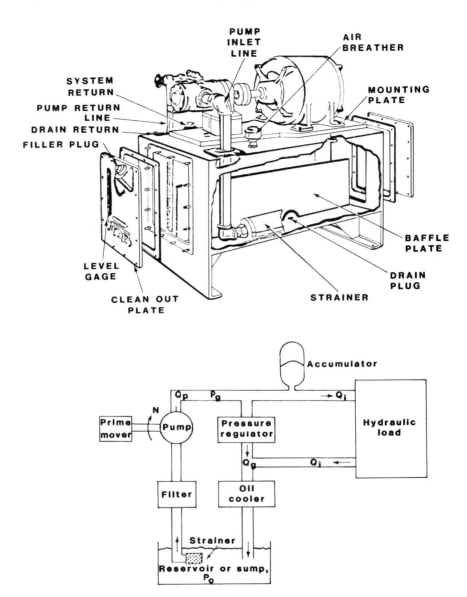

Figure 3–16. A hydraulic power supply. (a) Physical unit. (b) Flow diagram of the hydraulic unit. (Adapted from Herb Merritt, *Hydraulic Control Systems*. Reprinted by permission of John Wiley & Sons, Inc.)

Operation of Industrial Robots | **49**

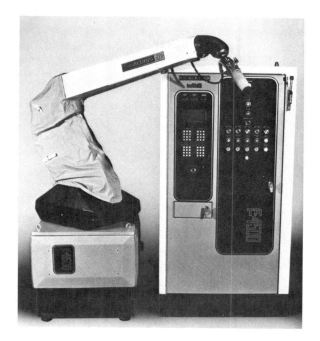

Figure 3–17. The DeVilbiss/Trallfa TR-4500 spray-finishing robot uses a hydraulic power system. (Courtesy of DeVilbiss Co., Toledo, Ohio.)

The word *servo* refers to a continuous-position controlling device. A nonservo robot may use a limit switch to indicate that the robot has reached the desired or end position. A servo system provides continuous positioning information along the path of the robot's movement.

Open and closed loop may also be used in reference to the overall robot manipulator. Five different classes of robot controllers are shown in the block diagrams in Figure 3–20. The first, shown in Figure 3–20a, is an open-loop axis controller that uses no feedback signals. The second, shown in Figure 3–20b, is called a nonservo controller and receives only on-off feedback signals. This type is in common use today and consists of an overall open-loop control, but local closed-loop joint controls. The third type, shown in Figure 3–20c, uses servo motors and feedback for each axis and is called a servo controller. This popular type uses local environment sensors that sense position and velocity to provide feedback information. The fourth type, shown in Figure 3–20d, is a more intelligent control with sensors in both the local and global environments to provide an overall feedback control. This type of control has not yet been implemented on an industrial robot. The fifth type, shown in Figure 3–20e, is an intelligent closed-loop control in which both local and global sensors are used to create and modify the strategy of the robot. This type is also currently in the research stage of development.

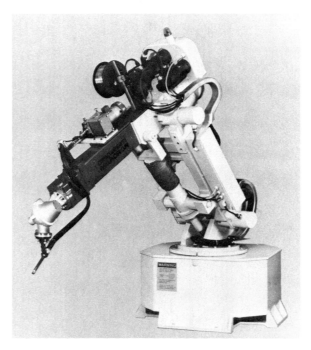

Figure 3–18. The Cincinnati Milacron T3 726 electric-drive robot with arc-welding torch. (Courtesy of Cincinnati Milacron.)

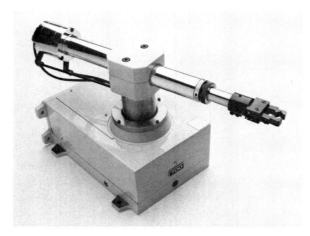

Figure 3–19. The Seiko Model 700 pneumatically powered industrial robot. (Courtesy of Seiko Instruments USA, Inc., Torrance, California.)

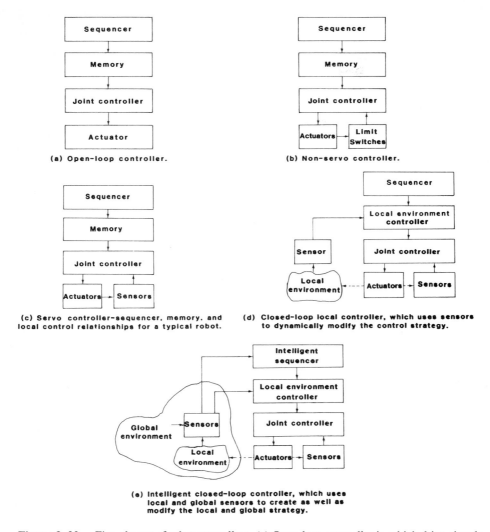

Figure 3–20. Five classes of robot controllers. (a) Open-loop controller in which drive signals are sent to actuators but no feedback is used. An example of this type of control is the stepper motor actuators used on the Microbot MiniMover. Servo motors may also be used in this manner. (b) Bang-bang, off-on, or nonservo control in which drive signals are sent to the actuators but a return signal is sent back to the drive motors when the desired position is reached. Nonservo robots, such as the Seiko, use this simple form of feedback. (c) Servo control in which drive signals sent to the actuator are compared with measured signals from the axis to control the motion of the axis. (d) Closed-loop local controller that controls all axes in a coordinated manner. This type of control has not yet been implemented in an industrial robot because of computational complexity. (e) Intelligent closed-loop control in which the local and global sensors are used to create and modify the robot strategy and motion. (Adapted with permission from *Engineering Intelligent Systems: Concepts, Theory and Applications,* by Robert M. Glorioso and Fernando C. Colon Osorio, Jr., Copyright by Digital Press/Digital Equipment Corp., Bedford, Massachusetts, 1980.)

Nonservo Robot Operation

The operation of a nonservo robot might be as follows for the control of a single axis. A controller is used to initiate signals to the control valve for the axis motion. The control valve opens, admitting air or oil to the actuator, which would be either pneumatic or hydraulic. The actuator starts the robot axis moving. The valve remains open, and the member continues to move until it is physically restrained by contact with an end stop. A limit switch placed at the end stop is used to signal the end of travel back to the controller, which then commands the control valve to close. If the controller is a sequencer or device capable of sending a sequence of control signals, it then indexes to the next step, and the controller again provides an output signal. These signals may go to the actuators on the robot manipulator or to external devices, such as the gripper. The process is repeated until the entire sequence of steps is completed.

Some features of this design are that the manipulator's members move until the limits of travel or end stops are reached. The number of stopping positions for each axis is at least two, providing the starting and stopping positions. It is possible to have intermediate stopping positions; however, there is a practical limit to the number of these that can be installed. This type of robot is thus limited in the number of positions in space that can be reached. If a six-axis robot has only start and end stops for each axis, then 2^6, or 64, positions can be reached. To soften the shock upon reaching a stop, a shock absorber or additional valving may be used to provide deceleration. The sequencer may be programmed and conditionally modified through the use of external sensors. However, the sequence for this class of robots is usually restricted to the performance of a single program, such as that required to pick up an object at a fixed location and place it at a given location.

Several characteristics make the nonservo robot ideal for certain tasks. One such characteristic is the relatively high speeds achievable, since a control valve can provide the full flow of air or oil to the actuator. Manual speed adjustment may be provided by regulating this flow. These robots are also relatively low in cost, simple to operate, set up, and maintain, offer excellent repeatability, and have high reliability. They have been mainly used in materials-handling tasks for investment casting, die casting, conveyor unloading, palletizing, multiple parts handling, machine loading, and injection molding. A typical nonservo robot is shown in Figure 3–21. The dual-gripper design would permit the robot to load and unload a machine tool without having to rotate completely. The control loop of a nonservo robot is shown in Figure 3–22a. Note that the nonservo robot is limited to situations requiring little adaptability.

As an example of the operation of machine loading and unloading, let's consider the Prab Model FA robot installation at Eaton Corporation in Marshall, Michigan. In this application, the part to be machined is a 20-pound malleable iron casting used for a locking differential for a three-quarter-ton truck. It is a back-breaking job for humans to move 12 tons of parts per shift, and the resulting operator fatigue can lead to a decline in productivity and possible quality problems. It was decided to group three drilling and boring machines into a compact work cell designed to include a centrally located Prab Model FA cylindrical coordinate robot.

Figure 3–21. A nonservo controlled Prab robot, which is effectively used for many pick-and-place operations and is characterized by excellent repeatability and modest cost. Two arms, capable of working independently, qualify the new Prab Model 6200 robot for high-speed parts-transfer jobs, especially in the metal stamping press room. With five to nine axes of motion and ±0.008 inch repeatability, the Prab Model 6200 handles payloads weighing from a few ounces to 70 pounds with end-of-arm grippers. (Courtesy of Prab Robots, Inc., Kalamazoo, Michigan.)

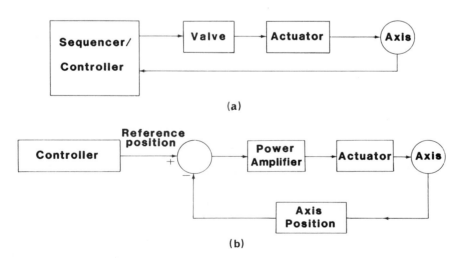

Figure 3–22. (a) The control loop for a nonservo robot. The controller or sequencer sends a drive signal to a valve that drives the actuator. When the axis reaches the desired position, a signal is sent back to close the valve. A characteristic of this type of control is that the robot can stop only at positions in which stops have been set. (b) The control loop of a servo robot. The controller sends a reference desired position. This desired position is compared with the actual position as measured by sensors, such as a shaft encoder and tachometer, where the error signal is used to drive the actuator through the power amplifier. When the error signal is zero, the actuator stops.

This example of a work cell is illustrative of the type of processes that may be included in the factory of the future. The operation of the cell is as follows. A pair of bell-shaped housings arrive at the work cell on an indexing conveyor precisely oriented on a fixtured pallet for the robot to pick up. Photoelectric cells are used to communicate to the robot that the parts are in position and ready to pick up. The machines are designed to accept two housings at a time. Therefore, the robot's gripper is designed to pick up and simultaneously move two parts through the three machines. The machines perform a drill-bore-drill operation. The first machine performs a drilling operation, and the second performs a boring operation. Both these machines require the parts to be oriented end to end, with the housing sides exposed for machining. The third machine performs a drilling operation requiring the parts to be oriented side by side. The dual grippers mounted on a rotary cylinder attached to the robot arm are programmed to accomplish this positioning. Following this last operation, the robot moves the parts to an unloading fixture on a simple outgoing belt conveyer. The robot is sequenced from unloading the third machine back to the incoming conveyor because the machine fixtures must be emptied before another pair of parts can be loaded. The robot performs a sequence of 49 steps to complete this cycle.

Servo Operation

A basic servo-controlled system receives its reference position signal from the sequence controller. The axis position measurement device also provides a feedback signal proportional to its current location. The difference between the desired and current position is called the error signal. This signal is converted to the proper form and applied to the actuator. If there is a large difference, a large signal is applied to the actuator and it moves quickly. If the error signal is zero, no signal is applied to the actuator, since it is at the desired location. With proper design, the action of this feedback is very smooth and reliable.

The operation of a more modern controller will now be described. Upon start of execution, the controller addresses the memory location of the first command position and also reads the actual position of the axis from the position-measuring device. The desired and actual position signals are subtracted to form an error signal. The error signal is then amplified and converted to a velocity signal. The actual velocity signal is read from a velocity-measuring device, such as a tachometer. The difference between the desired and actual velocities is used as another error signal. This velocity error is fed to a compensation network, which serves to keep the controlled motion stable. The output of this network is amplified and used to control the actuating device that moves the robot arm. The position and velocity feedback signals are linked directly to the robot axis. Industrial examples of the use of servo-controlled robots are shown in Figure 3-23.

As the actuators move the manipulator's axis, the feedback signals are compared with the desired position data, generating new error signals that are used to command the robot. This process continues until the error signals are effectively reduced to zero, and the axes come to rest at the desired position. The controller then addresses the next memory location and responds appropriately until the entire sequence or program has

Operation of Industrial Robots | 55

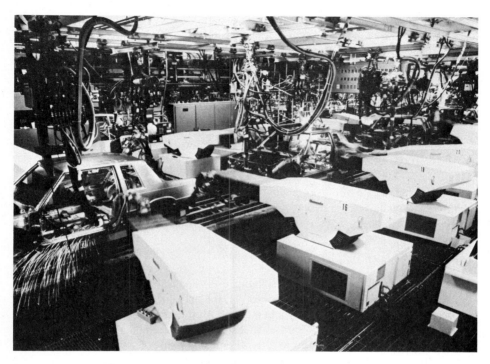

Figure 3–23. Examples of servo-controlled Unimate robots. A characteristic of this type of control is that the robot can be commanded to stop at any point in its work volume. (Courtesy of Joseph F. Engelberger.)

been executed. A simple control loop for a servo-controlled robot is shown in Figure 3–22b.

One of the main features of the servo-controlled robot is its versatility. It can move to any point within its limit of travel. It is also possible to control the velocity, acceleration, and deceleration between program points. With many systems, one can specify the travel velocity between points, which permits dexterous movements. The repeatability can be varied by changing the magnitude of the error signal, which is considered "zero." This feature may be used to permit the robot to round off a path between control points. Servo control systems have been used for hydraulic-, electric-, and, more recently, pneumatic-powered systems.

Programming may be accomplished either by a teach pendant, which permits the manual insertion of program points, or by external control for off-line programming. Either the output of the feedback position devices or the location of the end effector may be stored in the memory of the controller computer. The memory capacity of this computer is usually sufficient to store thousands of program points.

The characteristics of a servo-controlled robot may be observed in the smooth motions with control of speed and acceleration. This permits the controlled movement of heavy loads or delicate operations for sophisticated tasks. Since the servo robot can be positioned at any point within its working envelope, it has maximum resolution. Furthermore, most computer controllers permit the storage of main programs, as well as macros or subroutines, and permit program transfer based on tool conditions or external signals. Because of their complexity, servo-controlled robots may be more expensive than nonservo-controlled robots; however, the greater flexibility permits them to accomplish a greater variety of tasks.

3.3 Methods of Motion Control

The control of the path of movement of the robot can be accomplished in several ways. Four popular methods are called continuous path, point-to-point, joint interpolated, and controlled path motion. These will now be described.

Continuous Path Motion

Continuous path control is illustrated in Figure 3–24 and is the most popular method for spray-painting robots. To program this motion, the operator, either by directly moving the robot or by using a teaching arm or pendant, leads the robot through the desired path. The controller records the robot position at a fixed time increment. The time increment may be variable within a range from 5 to 80 points per second. If the operator moves slowly, the recorded points will be close together, and if fast motion is used, widely spaced points will result. Some editing capabilities may also be available to permit correction of errors. Programming the continuous path motion sometimes requires considerable operator skill. It has been recommended, for instance, that the programming be done by the best operator on his or her best day.

The actual path of the robot between the control points may not be a straight line, since it is a function of the actuator response action in the fixed time increments. The programmed points must be stored, and cassette tapes or floppy disks may be used to store several thousand points. An example of this kind of painting robot was the Japanese robot at the 1982 World's Fair exhibit, which used different programs to paint various figures for sightseers to take home as souvenirs. An interesting point about this robot is that the robot arm vibrated at its stop points, which made the pictures come out differently each time. Such vibration would be undesirable for extreme precision.

Point-to-Point Operation

The point-to-point programming method consists of moving the robot arm to each point using joysticks or push buttons to move the axes individually during teaching. At the desired location, a program button is pushed that stores the position information of that point, as shown in Figure 3–25. Points can be inserted at either closely or widely spaced intervals. In the standard point-to-point control, in playback or automated operation, all

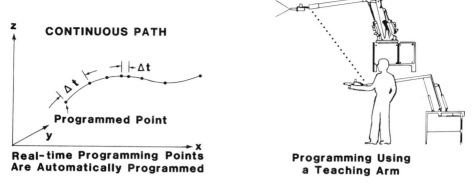

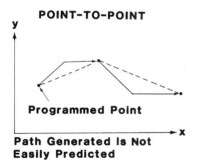

Figure 3–24. Continuous path motion as used, for example, on a spray-painting robot. During the programming stage, actuator control points are read and stored continuously or at discrete time intervals. If points are stored at discrete time intervals, the distance between points may vary. Upon playback the robot moves through the stored points in a point-to-point manner. (Adapted from R. E. Hohn, "Application Flexibility of a Computer Controlled Industrial Robot," MR76-603, 1976. Reprinted by permission of the Society of Manufacturing Engineers, Dearborn, Michigan.)

Figure 3–25. Point-to-point control. During programming, the actuator positions at selected points are stored. Upon playback, the robot moves through the programmed points. Motion between the points may be made in a coordinated manner by adjusting the velocity so that all axis motions end at the same time. (Adapted from R. E. Hohn, "Application Flexibility of a Computer Controlled Industrial Robot," MR76-603, 1976. Reprinted by permission of the Society of Manufacturing Engineers, Dearborn, Michigan.)

axes move at the maximum rate in a rather uncoordinated manner called the "race" mode. Whichever axis has the smallest distance to move will reach its position first, then wait for the others. This results in a path between points that cannot be easily predicted. Consider the following table, which is an example of the joint motions for a rotary axis robot with a fixed velocity of 10 degrees per second.

Robot Components and Operation

Joint	Amount of rotation (deg)	Speed (deg/sec)	Time (sec)
1	40	10	4
2	80	10	8
3	20	10	2
4	60	10	6
5	40	10	4
6	100	10	10

Note that joint 3 will reach its desired position in only 2 seconds but joint 6 will not be at the desired location for 10 seconds. The resulting motion may appear somewhat awkward.

Joint Interpolated Motion

Modified point-to-point or joint interpolated motion is designed to provide smoother motion. The points are programmed in a manner similar to the previous method. However, on automated playback, the velocity of each axis is adjusted. In this "proportional" mode, the axis that has the furthest to go, goes at maximum velocity. The other axes go at proportionally slower speeds. The following table clarifies this. Note that all axes reach their final position at the same time. The action is more coordinated than in the "race" mode; however, it is still not easy to predict the path.

Joint	Amount of rotation (deg)	Speed (deg/sec)	Time (sec)
1	40	4	10
2	80	8	10
3	20	2	10
4	60	6	10
5	40	4	10
6	100	10	10

Controlled Path Motion

Controlled path motion is still programmed at discrete points. However, the motion between points is a controlled path, such as a straight line, as shown in Figure 3–26. Intermediate points are computed so that a straight line path may be followed. This involves an internal computation between the commanded world coordinates of the robot and the joint angle coordinates. For example, the points programmed may be

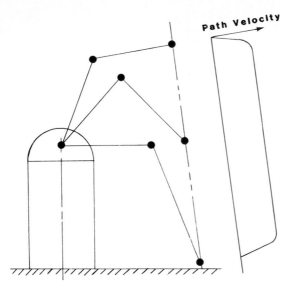

Two alternate ways of representing the path of a robot

Figure 3–26. Controlled path motion. This type of motion ensures that the robot follows a straight line between programmed points. An interpolation between programmed points is required to supply intermediate control points. This operation requires a computer capable of computing transformations between joint angles and hand position and orientation. The velocity and acceleration may also be controlled. (Adapted from R. E. Hohn, "Application Flexibility of a Computer Controlled Industrial Robot," MR76-603, 1976. Reprinted with permission of the Society of Manufacturing Engineers, Dearborn, Michigan.)

60 | Robot Components and Operation

stored in either joint angles or world coordinates. However, for interpolation of intermediate points, the simplest computation is done in world coordinates. The command signals must again be in joint coordinates. This leads to the coordinate transformations (Huston and Kelly, 1982). These computations may be made either off-line and stored for playback by the robot, or computed on-line. The on-line computation provides the greatest flexibility of robot control. A vivid demonstration of the straight-line motion is shown in Figure 3–27. A ruler may be placed along the path of the end tool in this illustration to show the precision.

3.4 Hierarchy of Control for Robots

The overall control of a robot is a complex problem consisting of strategy development, path planning, sensory information integration, position commands, and drive signals. One method of simplifying this problem is to break it into levels of a hierarchy. In the hierarchy, items or tasks that are grouped together at a low level remain grouped through the higher levels. A family hierarchy with parents and children is a familiar example. A web or network in which there is more than one path to a particular mode is a more complex example.

The hierarchical control strategy for robots is described by Barbera (1977). He

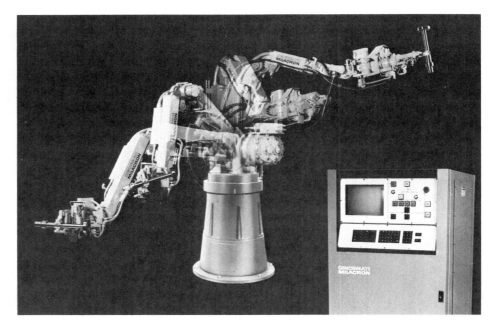

Figure 3–27. A Cincinnati Milacron T3 robot in a vivid demonstration of straight-line, controlled path motion. (Courtesy of Cincinnati Milacron.)

starts with a first-level control that receives joint position commands and sends drive signals to the robot's joint actuators. These first-level control loops often have feedback from position and velocity indicators to ensure that the joint motion is controlled and stable. In implementation, the first-level controller may be an analog, digital, or hybrid circuit. Stability, speed, and repeatability are key design specifications.

The second level control receives as input such commands as MOVE TO (X,Y,Z) and interprets this global command into individual joint position commands. These are then sent to the first-level control. If the command GRASP OBJECT is sent, then tactile or force feedback from the end effector may be sent to the second-level controller to verify that a grasp has been accomplished. Proximity sensors may also be used to alter the approach speed of the gripper. The interpretative nature of the second- and higher level controls often dictates a digital controller for implementation. The ability to interface with external sensors and interpret commands are characteristics of the second-level control.

The third-level control divides or parses higher level commands into individually required actions for the second level control. For example, the command MOVE TO (X,Y,Z) AT VELOCITY 30 INCHES/SECOND, DECELERATE AT 12 INCHES FROM OBJECT, GRASP UNTIL FORCE EQUALS 1 POUND must be divided into motion commands, positioning commands, and grasping commands. Feedback from joint positions and velocity sensors, proximity sensors, and tactile force sensors is needed to execute the command.

Higher levels of control may also be developed. Each higher level takes more complex task commands and higher level sensory information and develops commands for the lower level controllers. For example, in executing the previous command, the position and orientation of the object being grasped must be known. This information may be determined by a vision sensor. If more than one type of object is available, then the part identity must be established. Also, the relation of the part to the other objects and its surroundings must be determined. If the object is covered by other objects, then these obstructing objects must be removed before the part can be grasped.

Higher levels of control are needed for positioning commands. If there are obstacles in the environment, a strategy for avoiding them must be developed. If a path has been specified, it must be checked to see if it can actually be followed. If a path has not been specified, one must be determined. Even higher levels of control, which could take a single command like BUILD A CAR from a human and develop all the required lower level commands, can be envisioned; however, a great deal of research is needed before such a system can be implemented. An illustration of a hierarchy of control is shown in Figure 3–28.

3.5 Line Tracking with Industrial Robots

The assembly line is a control concept in existing and planned future factories. It characterizes the division of a task into subtasks, such as the material processing required to transform a given raw material into a finished product. The stationary

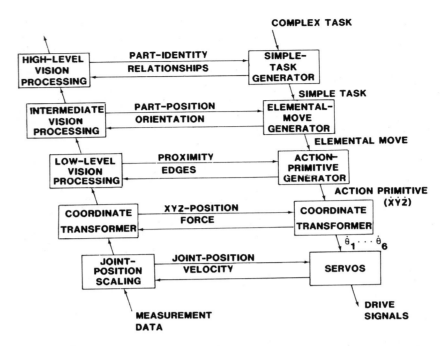

Figure 3–28. The hierarchy of control for servo robots. A complex task at the top of the hierarchy is divided into simple tasks, the simple tasks into elemental moves required to complete the task, and the elemental moves into primitive action commands in a world coordinate system. These coordinate commands are transformed into robot joint commands; the joint commands then become the drive signals for the robot actuators. The sensors required to accomplish each level of task accurately are shown on the left. Actuator joint positions and velocity are required by the servos. Overall measurement of position and force may be desired for the coordinate transformer. Low-level visual processing to determine position or proximity may be needed by the action generator. Higher level visual or tactile sensing may be required to determine the position and orientation of the workpiece. Very high level computer vision may be required to determine the position of obstacles, to permit path planning, and to react to unusual events in the environment. (Adapted from James S. Albus, *Brains, Behavior and Robotics,* 1981. Reprinted by permission of BYTE/McGraw-Hill.)

industrial robot is integrated into the assembly line by line tracking. Line tracking is the ability of a machine to carry out operations on parts mounted on a continually moving conveyor. Without line tracking, expensive, high-speed, and often heavy-duty shuttle systems must be installed and maintained. These shuttle systems take parts from the main conveyor to a location for processing, then return them to the main line. If the processing time is too long or the process robot too limited, a shuttle system is required. However, a robot with line-tracking capability can eliminate this nonproductive shuttle operation and reduce the part cycle time.

Industrial robots may be used for line tracking in two ways. In the moving-base line-tracking system, the robot is mounted on a traverse base. Generally, this method gives the robot a longer time to work on a given part, but requires an expensive traverse system. The second method is stationary-base line tracking. This method gives the robot access to a larger portion of the object and eliminates the need for the traverse system. Special considerations are required with line tracking. In the simplest case, a position resolver is attached to the conveyor and calibrated, and a part in the range switch is installed. Teach programming is then performed on a stationary part. This taught program can then be executed on the moving part, even at variable speeds.

3.6 Modular Robots

How is a robot built? Most industrial robots are custom-built to the specifications of the buyer. However, one other method is to use modular components and assemble the type of robot desired. Mack Corporation supplies nine basic components that can be combined to build a 6-degree-of-freedom Cartesian robot. The nine basic components are the grippers, adapters, x-axis, y-axis, and z-axis transporters, roll, pitch, and yaw rotators, and position incrementers, as shown in Figure 3–29.

The grippers are provided in four sizes in two- and three-finger configurations for

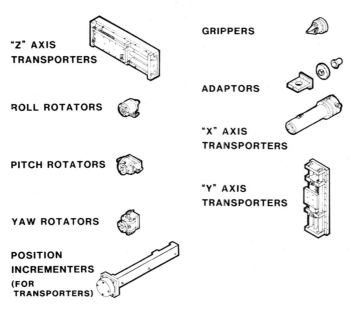

Figure 3–29. Basic components of the Mack Corporation modular robotic system. The components may be concatenated to form a variety of different Cartesian robot designs. These robot components are powered by pneumatic or electric sources. The maximum load capacity is 5 pounds. The resulting manipulator, when combined with a power system and controller, can make a complete robot system custom-designed to an application. (Courtesy of Mack Corporation, Flagstaff, Arizona.)

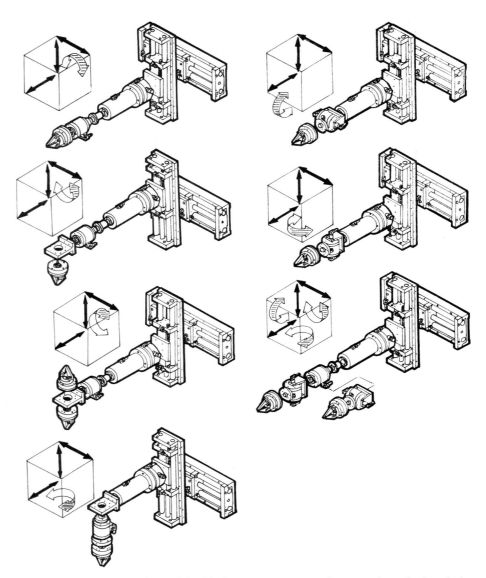

Figure 3-30. Combinations of the Mack components to produce a variety of robot designs. (Courtesy of Mack Corporation, Flagstaff, Arizona.)

external or internal gripping, as well as soft blank fingers that can be easily modified for special shapes. The smallest size is about twice as large as a thimble and the largest about the size of a human hand. All units operate on the principle of a double-acting cylinder controlled through a simple four-way valve circuit. Fluid pressure opens or closes the fingers. Maximum operating pressure is 150 pounds per square inch (psi) in either hydraulic or pneumatic service. However, most applications use plant air at 80 psi for a reliable source of fluid power. At 80 psi, a pinch force between 5 and 50 pounds may be developed, depending on the model size.

The gripper may be mounted in several orientations. Perhaps the simplest is to mount the gripper in line with the first desired motion. This mounting is facilitated by the various adapters. The transporters are the next elements to consider. The x-axis transporter is an air cylinder that provides straight-line motion. Special features include adjustable travel stops and pistons keyed against rotation. The x-axis transporter may also be combined with y- and z-axis transporters. These transporters may also be powered with plant air and may be controlled by simple air logic, or by programmable controllers or computers when more sophistication is required. Rotators are also key components in the modular system. The rotators are nonservo, air-operated, vane-actuated units with a choice of 90- or 180-degree rotations in roll, pitch, and yaw. Adjustable stops provide close control over the angular positions.

Many different designs can be constructed from the basic components, as shown in Figure 3–30. Combinations of the various components can provide a 6-degree-of-freedom, rectangular robot with a payload of 5 pounds and nonservo operation. Controllers are also provided. However, it should be noted that the modular robot components described here are for a fixed-sequence, nonservo positioner.

Questions

1. What is the number of degrees of freedom required to position a robot manipulator at any point in three-dimensional space?
2. What is the work volume of a cylindrical robot?
3. A computer controller is essential in industrial robots for obtaining what controlled path motion?
4. In the hierarchical control strategy for a servo industrial robot, what is the function of the first-level control?
5. What are the four commonly used industrial robot configurations?
6. In assembly-line tracking with industrial robots, two methods were discussed. What were they?

Intelligent Robot Programming

> Thus man is the most intelligent of all animals and so, also, hands are the instruments most suited to an intelligent animal. For it is not because he has hands that he is the most intelligent, as Anaxagoras says, but because he is the most intelligent that he has hands, as Aristotle says, judging most correctly.
>
> Galen, Greek physician
> (ca. 130–200), *On the
> Usefulness of the Parts
> of the Body* (Boorstin, 1983)

Industrial robots with smart sensors, such as vision, touch, and hearing, and machine intelligence promise to significantly increase the flexibility of robots within the next decade. According to the RIA *Worldwide Robotics Survey and Directory* (1983) there were 6301 robot installations in the United States at the end of 1982. Annual production of robots is expected to be 24,000 units by 1990. The total projected number of robots in use in the United States by 1990 is 100,000, and the Japanese expect to have 557,000 robots in use by then. Thus by 1990, several hundred thousand robots will be operating in the world, and there may well be several million by the year 2000. Many of these new installations will be using the new generation of intelligent robots that rely heavily on visual and other sensors to permit them to be intelligent in the sense that they can accommodate changes in their environment. A large number of machine and artificial intelligence techniques have been developed in the field of computer science that can now be applied to the manipulation of robots.

Intelligence is often demonstrated by game-playing skills. Chess, checkers, and Rubik's cube are just some of the games that can now be played by intelligent robots.

Sensors detect the environment. The environment in these games may be the locations of the chess or checkers pieces, or the color patterns on the cube. A computer then determines the best move, using a program based on artificial intelligence techniques, which lets it search through possible solutions. Then, the computer commands and controls the robot manipulator to carry out its series of motions to accomplish the action. A more practical intelligent application is in palletizing parcels of mixed size and weight, which requires a space-filling solution. Such tasks as part scheduling, complex assembly, and mobility also require intelligence.

4.1 Artificial Intelligence

Intelligence is a fundamental human characteristic and has many varying definitions, implied meanings, and levels of sophistication. Let us consider some of the ways intelligence may be applied to robots.

Robot intelligence permits a robot to adapt to various changes in its environment. The foundations of this ability are found in previous developments in the fields of artificial and machine intelligence. Since intelligence is difficult to quantify, let's start by reviewing the basic definition of human intelligence.

Intelligence is defined in Webster's dictionary several ways. We will consider the following:

1. The ability to learn or understand or to deal with new or trying situations
2. The ability to apply knowledge to manipulate one's environment

The first ability is much more difficult to accomplish than the second. In fact, we may distinguish between the studies in artificial intelligence (AI) and machine intelligence (MI) by these abilities. An attempt to implement the first ability into a machine may be called AI, and the implementation of the more modest second ability may be called MI. The goal of AI is to produce computer systems that imitate human performance in a wide variety of intelligent tasks. The goal of MI is to design a useful, adaptive, intelligent machine. These goals may be expanded.

The goals of AI include the following:

1. Finding new methods for extracting useful information from sensors
2. Developing methods for building, updating, and retaining information from a knowledge base
3. Inventing algorithms for utilizing the information stored in a knowledge base for making intelligent decisions
4. Finding improved methods for translating needs into a workable software system
5. Developing reusable software components that can expand toward an ultimate software system

AI systems require a knowledge base containing several types of information, such as that about objects, processes, reasonable goals, and hard-to-represent concepts about time, space, and causality. Knowledge representation, therefore, forms the foundation upon which much AI activity is based. Many new questions about knowledge representation have been raised in the past few years. For instance, what is the best way to structure the knowledge for ease of access, storage, and use? How can a set of rules for manipulating the specific elements in a knowledge base be devised so that implicit as well as explicit knowledge can be inferred? How is incomplete knowledge to be handled? What methods can be used for acquiring new knowledge? What is the best way to extract knowledge from a human expert? How can queries about the data base be answered? These and many other questions form an exciting basis for research in AI.

Artificial intelligence is of interest not only in understanding the human but also in providing a basis for building intelligent machines. Some of the topics included under the umbrella of AI include knowledge representation, pattern recognition, computer vision, reasoning, natural language understanding, and cognition. The scientific goal of understanding the nature of intelligence is as fascinating as Einstein's involvement with understanding the fundamental principles of nature; however, it is looking inside ourselves rather than outward toward the galaxies. Attempting to build intelligent robots is a practical goal, not only permitting us to develop some very useful machines, but also leading us to very basic questions about intelligence that must be answered to understand ourselves.

The most basic act of intelligence is simply making a decision, such as to eat or not to eat. This level of intelligence is found in all the forms of animal life and is also the basic process that led to the development of computers and robots. In fact, this level is so well integrated in the human that it is considered a rather oversimplification of what we mean by machine intelligence. Today's researchers in AI are more interested in higher levels of intelligence, such as reasoning or "common sense." Common sense is vital to human survival and behavior. However, it is so difficult to define and involves so much complex reasoning that it is often cited as a quality that even some humans lack. Another interesting problem relates to how reasoning ability or judgment works. Human reasoning is very advanced. We often make decisions based upon incomplete or partial information about a situation. For example, investments in the stock market are regularly made on the basis of incomplete information. Another aspect of intelligence is expert reasoning, or the ability to make judgments based upon information that is not easily stated or even consciously recalled. For example, a physician may be able to diagnose an illness in a very short time, perhaps less than a minute. However, if asked to describe the basis of the diagnosis, the physician may require hours to describe the information from the books, medical school courses, and case histories that formed the basis for the judgment. One approach to designing an expert machine is to couple an expert, such as the physician, with a computer scientist, and let them work together to develop a computer program that integrates the expert's knowledge into a computer system. Some examples of difficult problems that have been solved in this manner include medical diagnosis, oil exploration, geologic fault isolation, genetic experimentation, and visual recognition. One implementation of this approach involves writing a computer program that asks questions of the expert to develop a knowledge data base, then continues to form a relational model from the data base, and finally provides an expert program that can make judgements similar to those of the human expert. An example is a program that can write other computer programs. A set of questions that defines the problem and solution is asked by the machine and answered by the human. If sufficient information is given, a computer program to solve the problem can be developed. Other examples may be found in the area of computer-aided design (CAD). Here, a collection of design and analysis programs is made so that the designer simply specifies the parameters of the design and the computer analyzes the performance of the specified system. The results of the analysis are presented to the designer so that they may be compared to the original design goals. The design may then be analyzed and altered repeatedly to provide a final design. As an example of the need for this iterative process, consider the drawing of an "impossible" figure shown in Figure 4–1a. On first glance the drawing seems reasonable. However, upon checking the consistency of the drawing, it can be clearly determined that the figure cannot be built although the attempts shown in Figure 4–1b are as close as possible.

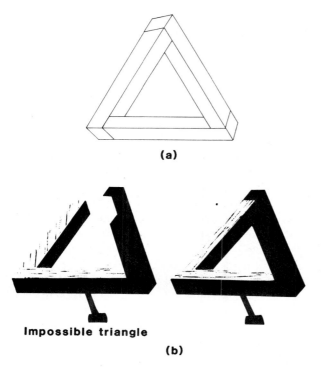

Figure 4–1. An impossible figure. (a) Drawing that appears reasonable. (b) A realization that closely resembles the design. (Adapted with permission from *Engineering Intelligent Systems: Concepts, Theory, and Applications,* by Robert M. Glorioso and Fernando C. Colon Osorio, Jr. Copyright Digital Press/Digital Equipment Corp, Bedford, Massachusetts, 1980.)

Learning from experience is another characteristic of human intelligence. One approach to machine intelligence to learn in this manner provides the machine with prototypes or examples of the desired pattern to be recognized. The machine extracts features from the prototypes and develops a decision rule based upon these features. Learning with a teacher describes the method in which a human provides the prototypes in a labeled manner so that the features of a particular class may be grouped together and then compared to the features from another pattern class. This approach has been successfully used in many applications but may require a large effort in the collection of prototypes from each pattern class. Learning without a teacher may also be accomplished by simply presenting the machine with an assortment of patterns. The machine must determine the number of classes and characteristic patterns that are used to make decisions on new test samples. This process may lead to new discoveries of pattern classes; however, long training periods may be required. Interestingly, it is common to require as many as 20 years of formal education with teachers for humans to learn from the collective experience of humankind and develop the ability to work independently. Only a few intelligent machines or programs have been developed over a 10-year period. Perhaps a longer view of machine learning is required of machines just as it is for humans.

Understanding natural language is one area that has been studied for about 20 years, with some considerable progress. Word recognition and synthesis can now be demonstrated on microprocessors in a rather limited but effective manner. Since this is a very effective form of human communication, more research can be expected in this area.

Performing appropriate actions in unusual circumstances is another characteristic of high-level intelligence. For limited action situations, such as games, the universe of all possible solutions can be represented by a tree structure to permit a machine to make a response to any possible situation by searching the tree branches in a clever manner. A clever search is required if the number of branches is too large to be searched exhaustively. We are still far from achieving this level of intelligence in the general situation. Although one can conceptualize searching all possible paths and deleting those that do not lead to a desired goal, it is possible that the computing time required to do this could run into hundreds of years, such as would be required for a game like chess.

The concepts of generalization and specialization are important to mention in this discussion of intelligence. We are constantly striving to generalize our knowledge about ourselves and the world. This process is endless. The knowledge that has been accumulated permits us to specify and engineer solutions that may be useful although limited in scope. In only a few instances have we developed universal solutions. The computer, which performs a finite number of operations to permit an infinite number of calculations, and the robot, which permits a finite number of degrees of manipulative freedom to provide an infinite number of possible motions, are noteworthy examples. The modern robot, which combines these two capabilities, offers unlimited application. Even when we do not have a general solution, a specialization of the solution may be useful. Examples are machine recognition of printed characters and limited speed recognition.

Perceiving the world around us and responding to the changes is a relatively modest form of intelligence. For example, using an umbrella on a rainy day is a rather mundane, or commonsense reaction. This second sense of the definition of intelligence is very closely related to the form of intelligence we are trying to build into today's "intelligent robot." Interestingly, this is not such a simple task. In addition to the computer and manipulator, the intelligent robot requires sensors to permit it to understand the environment and provide a basis for reacting to changes. Sensors require design, processing, and simple implementation to be useful. Robots without sensors cannot possibly be intelligent. Robots with sensors may or may not appear to be intelligent, depending upon their program, but they do have the capability to be intelligent.

Artificial intelligence has now been studied seriously for the past 25 years. For example, Professor Marvin Minsky used robots, vision systems, and computers in his work at MIT, as shown in Figure 4–2. In the early days, direct analogies between the human brain and the elements of a computer were made. The following are examples of such analogies:

Brain.....................................Computer
Knowledge............................Data

Figure 4–2. Studies in intelligent robotics conducted by Professor Marvin Minsky at the Massachusetts Institute of Technology. Note the vision system, the robot, and an artificial intelligence program manipulating objects from the blocks world. (Courtesy of the Massachusetts Institute of Technology Museum, Cambridge, Massachusetts.)

Memory	Storage
Judgments	Decision making
Stream of consciousness	Program execution
Reasoning	Heuristic search
Learning	Pattern recognition
Psychology	Artificial intelligence
Vision	Image processing
Hearing	Language understanding
Speech	Voice synthesis
Movements	Robots
Common sense	Knowledge reasoning

Many consider such analogies naive, and we would have to concede. The differences between a brain and a computer are perhaps greater than the similarities. However, a great deal has been learned about human intelligence by attempting to describe intelligent activities and phenomena in sufficient detail to permit one to write a program to simulate it on a computer. However, several computer programs have been written that can demonstrate rather conclusively that the computer can outperform the human in many cases, such as at playing games. How do human and computer storage capacities compare in terms of understanding the world we live in? As Carl Sagan (1979) states, "we do not have the storage capacity either in the human brain or in our largest computers to understand a barely visible grain of salt." Let's explore this statement.

Salt is sodium chloride, or NaCl. The atomic weight of salt (Na) is 22.99 and that of chlorine (Cl) is 35.45. Therefore, the molecular weight of NaCl is 58.44. One molecular weight of NaCl weighs 58.44 grams and contains 6.02×10^{23} molecules. A barely visible grain of salt weighs approximately 1 microgram. The number of molecules in a microgram of NaCl is (6.02×10^{23} molecules $\times 10^{-6}$ grams)/58.44 grams, which is approximately 10^{+14}. Therefore, to describe the three-dimensional position of each molecule requires 3×10^{14} numbers. If each number is represented by a 7-bit binary number, then the storage capacity required to understand this barely visible grain of salt is about 2×10^{15} bits.

The human brain is known to have about 10^{11} neurons. If we assume that storage capacity is represented by a dendrite connection between neurons and that each neuron is connected to about 1000 others, then the total storage capacity of the human brain is about 10^{14} bits of information. Thus, we see that the storage capacity of the brain is less than 5 percent of that required to specify the position of the molecules in a microgram of salt.

Rather than feel limited by this comparison, however, let us compare the storage capacity of the human brain to some other forms of information storage. A book of 500 pages with 500 words per page and 5 characters per word contains about 1,250,000 characters. If each character is represented by a 7-bit code, such as the American Standard Code for Information Interchange (ASCII), then the book would contain 8,750,000 bits of information. The human brain could store the contents of more than 10 million such books.

The on-line storage memory of a supercomputer, such as the CRAY, is about 8 million characters, or 56 million bits. The storage capacity of a large magnetic disk is about 300 million characters, or 2.1 billion bits, which is still much less than the human brain. The *Encyclopedia Brittanica* has about 12,500 million characters, or 87.5 billion bits. All the written information in the National Archives has been estimated at 12,500,000 million characters, or 8.75×10^{13} bits. Even this large amount of information would not fill the 10^{14} capacity of the human brain.

The storage capacity of a robot controller may only be 64,000 characters, or about 448,000 bits. This is certainly a modest amount when compared with a human's capacity of 100 trillion bits.

4.2 Machine Intelligence

The concept of a universal machine, which has not only the intelligence capacities of the general-purpose computer, but also the manipulative capacity of a general-purpose robot, has such far-reaching applications, consequences, and perhaps pitfalls that it is difficult to describe them. The number of actual implementations of intelligent robots is still rather limited in both quantity and quality. However, the need for such systems in industrial situations is clear.

Sensors for sight, touch, hearing, sound, taste, and smell are very important to humans because they permit us to accommodate to changes in our environment. Similar sensors are useful to the intelligent robot. However, extensions of the common human senses developed over thousands of years of science can also be added to the robot.

Tools of great variety would also be needed by the universal machine. Grippers with a variety of fingers, ranging from a simple hook that could pick up a bucket to a two-fingered gripper that could pick up an object, to a three-fingered gripper that could roll a pencil in its fingers, to a multifingered manipulator that could throw a perfect football pass with the appropriate spin, could be used. A dexterous hand similar to the human hand being studied by NASA for future space applications is shown in Figure 4–3. Such tools as screwdrivers, impact wrenches, welding devices, spray painters, and even calligraphic painting brushes have already been developed for robots. The list of tools or end effectors for the universal manipulator is also quite long. Is there a universal set of tools?

Robots capable of storing and performing a fixed number of operations can be made to perform a great variety of motions. Using a control device called a "teach pendant," which permits the robot to be positioned in space and programmed, a human trainer can instruct the computer-controlled robot, which will be able to perform the sequence taught with a far higher degree of repeatability than could its human trainer. Furthermore, by recalling the sequence from memory, the robot can perform the task repetitively and consistently. However, if something in the robot's work environment changes—say, the part the robot is to pick up is missing—then the robot will appear to be "dumb," since it must still follow those same, programmed motions just as if the part were present. To appear intelligent, the robot must be able to respond to changes in its environment: the robot needs intelligence. This ability to adapt has been crucial in the evolution and survival of humans and will be essential in the development of versatile

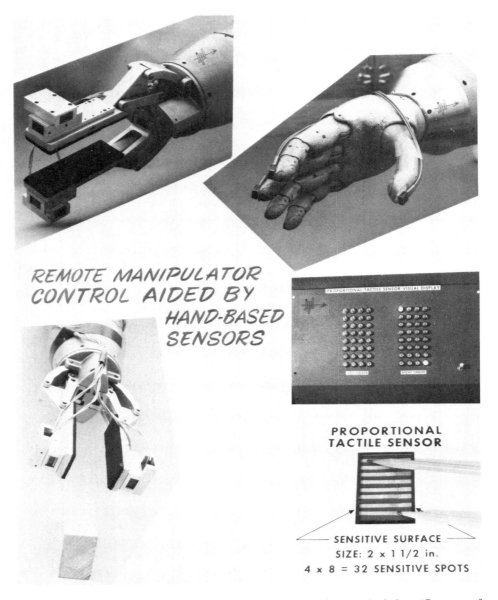

Figure 4–3 Dexterous hand being developed for versatile object manipulation. (Courtesy of JPL/NASA.)

robots. If the robot can sense changes in its environment—by using sensory perception—then it will have a better chance to solve problems for itself.

Robot Checkers Player

Equipping the robot with this adaptability involves the use of sensors and machine intelligence. Sensors are used to determine how the environment has changed, and machine intelligence is used to determine how the robot should respond to the changes.

To illustrate the steps involved in an intelligent robot, let's consider the cyclopean checkers player developed at the University of Tennessee (Hall et al., 1982). Since checkers is a well-known game, we won't go over the rules, but just the sequence of steps in the game shown in Figure 4–4. Let the robot move first. The robot arm moves out, and the gripper closes on one of its front pieces. It moves the piece forward. It is now the human's turn to move. Since the human has four possible checkers to move, the intelligent robot must sense the changes in the environment to determine where and when the human moves his checker. In the cyclopean system, this sensing is accomplished with a small, solid-state camera that monitors the board 30 times each second. It first stores an image of the board after the robot arm has made a move and has positioned itself out of the camera's view. When the human reaches to move his checker, the image in the camera changes, since the human's hand is in the field of view. This change can be detected and used to signal when the human starts and ends his move. After the human has removed his hand, a second image is taken and compared with the first. The first reference image is subtracted from the second image, which eliminates all information that did not change between the two images. Where the piece was and where it ends up are thus easily detected. At this point, the image-processing and pattern-recognition processes have extracted the relevant information from the scene. The robot now knows when and where the human has moved, and will next determine its response. The robot must now use machine intelligence to analyze the possible valid moves and to select the "best" move. The number of possible responses and the subsequent number of games are so large that an exhaustive search of all possible combinations is not feasible. Instead, a method or algorithm must be used that will search only a few of the possibilities and select a "good" move. Once the good move is determined, the robot simply reaches out and moves its piece to the desired position. The cyclopean robot plays a very natural-appearing game of checkers, even though it is implemented with microprocessors to control the camera and robot. It can accommodate multiple jumps, crown pieces, and remove jumped pieces from the board. It can even speak; for example, it may ask its human partner to wait until it finishes a move or to inform him that an illegal move has been made. It illustrates the key ingredients of any intelligent robot system—sensors, computers, and robot manipulators.

A Robot Solution to Rubik's Cube

If checkers is not challenging enough a game to convince you that intelligent robots can be "smart," then let's consider another game, solving Rubik's cube. A compact, self-contained robot called Cubot has been built for solving the three-dimensional

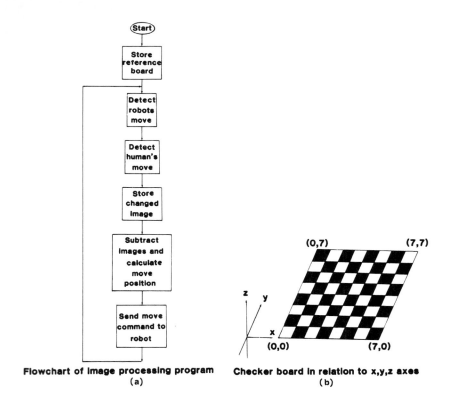

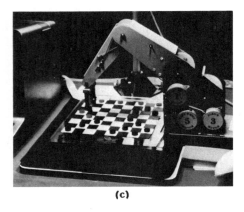

Figure 4-4. Intelligent robot vision system designed to play the game of checkers. (a) A flowchart of the image-processing program. A reference image is first stored. The robot's move is detected from the image. The human's move is then detected. A new image of the altered positions is recorded, then subtracted from the reference image to determine the human's move. The detected positions are then sent to the robot, which uses a checkers program to determine the appropriate move for the robot. (b) The checkerboard in relation to the global coordinate axes. (c) The robot manipulator moving a game piece. (d) The overall system with camera, robot, and checkerboard and Mr. John Lesac, one of the designers.

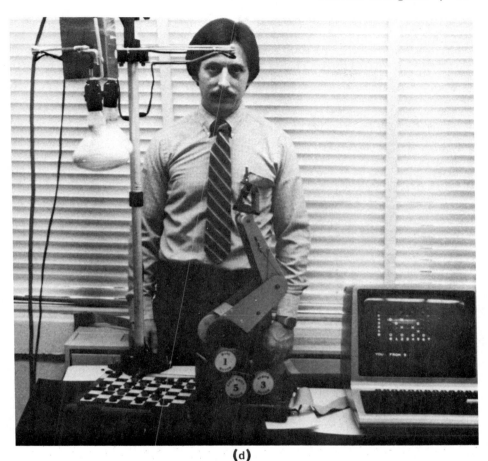

(d)

Figure 4-4 (continued)

Rubik's cube puzzle. The system, shown in Figure 4-5, was built by researchers at Battelle Pacific Northwestern Laboratories to demonstrate the capabilities of intelligent robots (Reich et al., 1983).

Erno Rubik's cube has six surfaces, each divided into nine facets that can be manipulated to produce a solid color for each face. Rubik originally intended his invention to be used by his students as an exercise in spatial thinking at the School of Commercial Artists in Budapest, Hungary. The mathematical implications of his invention fall into the branch of rather advanced mathematics called group theory. The number of positions that can be achieved by turning faces of the cube is about 4.3×10^{19}. It would take a computer over 1 million years just to go through all the possible combinations of all the possible positions of the cube. In light of this, it is a tribute to human ingenuity that even children have solved the cube in less than 1 minute.

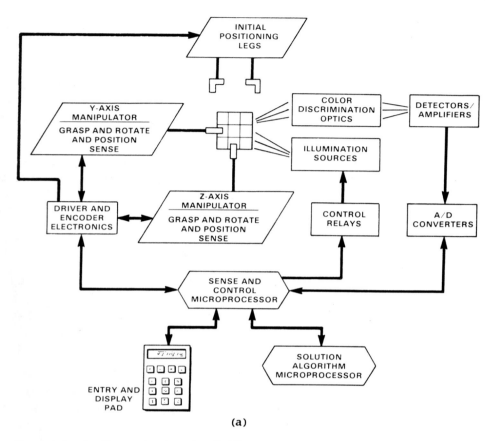

Figure 4–5. Intelligent robot that solves Rubik's cube. (a) Cubot system block diagram showing the components of the mechanical, optical, and electronic systems. (b) Actual system of Cubot. (Courtesy of Battelle Memorial Institute, Pacific Northwest Laboratories.)

Computer algorithms can determine a solution sequence in less than 1 second using AI techniques. However, the robotic solution also consists of determining the initial position and physically manipulating the cube to its solution. Students at the University of Illinois recently completed the first robot solution to the cube. The robot solution described here is a bit more sophisticated than the first solution and solves the cube without human intervention by using sensors, a computer, and a special robot.

The Cubot has three major subsystems. The electro-optical system illuminates the cube and provides a color readout of the cube faces. The microprocessor interprets these sensed data, computes a solution, and formulates an instruction sequence for the robot manipulator. The manipulators and grippers then rotate the cube faces based upon the solution sequence. Tests of the system indicate that it can solve the most scrambled cube in less than 3 minutes. Although it is not as fast as humans, the Cubot is a demonstration of the type of intelligent robot technology that can surely be called "smart."

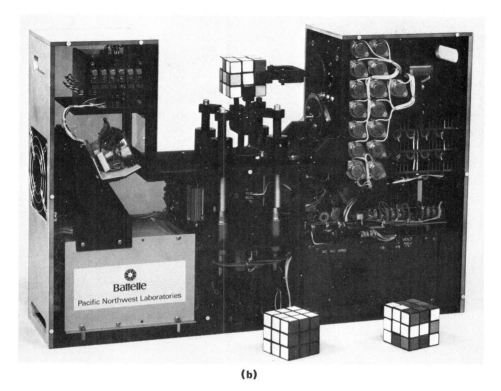

(b)

Figure 4–5 (continued)

As we have seen, cameras, visual point sensors, and tactile touch sensors may be used to sense the environment. In general, we may divide sensors into two categories, those that require contact with the surface and noncontact sensors. The camera is popular because it is a noncontact method of measurement and because it gives a set of measurements over a spatial region and not just at a point. It is natural to use a camera to guide a robot to a position, but it may be necessary to use a contact sensor to perhaps pick up a part at the position. In the cyclopean checkers player, a force sensor in the hand is used to check that the robot actually picked up a piece. In other situations, such as fitting a chess piece into a hole, a contact sensor is used to determine the location of the piece.

4.3 Voice Control of Robots

The use of speech recognition for the control of a robot provides a desirable method for human-machine interaction. Speech-recognition equipment is used in many applications, such as extending the capabilities of the physically handicapped. There are now a number of cases of people running complete offices from their homes, including typing, telephoning, and environmental control, such as lights and windows, using speech control. The use of speech control of a robot manipulator should provide a valuable extension in the ease of use of robots. In 1983, at the Automan exhibition in Birmingham, England, a Cincinnati Milacron T3 726 robot was demonstrated, that was controlled by a voice. Visitors would tell their initials to the robot, and the robot would engrave those initials on a glass paperweight submerged in water.

There are now about 250 different speech-recognition systems available worldwide. Recognition capabilities are limited from about 32 to 256 words. A training mode is used to permit the system to adjust to one or several speakers. Recognition rates of 99 percent have been reported. Representative companies include Verbex of Bedford, Massachusetts, a Division of Exxon Enterprises, Votan of Fremont, California, and Scott Instruments of Denton, Texas. This area appears quite rich for further developments.

4.4 Programming a Robot

The bridge between the robot, its sensors, and its application is the human programmer. A human must write the program, or at least write the program that writes the program to be used. The easier it is to develop or modify the program in a robot, the easier it is to solve problems for a given application. The need for ease of programming is great in giving the intelligent robot versatility. Of course, in many applications, an industrial robot will be installed and can run the same program for years. However, if a robot is to be in a work cell making small batches of products or if product changes are made periodically, then new or modified programs will be required almost continuously. Since the robot is designed to be versatile, the software should also be versatile.

Certain elements of the robot controller must be written in very efficient languages for speed. The basic servo control loops and drivers for sensors fall into this category. However, perhaps just as important as speed of operation is ease of use. This is why most manufacturers also supply a robot programming language and operating system support for the robot controllers for the software changes required to make the robot truly versatile.

Even if each manufacturer supplied a language to make its robots easy to use, we encounter a dilemma: there are almost as many robot languages as there are robots. One consequence of this is, if a manufacturer attempts to use more than one brand of robot in an application, the technical staff must learn more than one language. This is not an impossible task for the trained computer engineer or scientist, but it can be a burden to operators and technicians. One practical solution to this problem is for the company to specify a given type of controller and interface when purchasing a variety of robots. However, this is not the best solution.

If we look at the general computer world, the same phenomena occurred and are still going on. Each computer manufacturer must develop an assembly language unique to the architecture of its computer. However, groups of users quickly discovered that they were expending a good deal of redundant effort in developing assembly language subroutines to multiply, divide, compute trigonometric functions, and solve linear systems of equations. The invention of such high-level languages as FORTRAN for scientific users, COBOL for business users, BASIC for beginning users, ALGOL for mathematics, LISP for symbol manipulation, PROLOG for logic programming, and more recent structured languages, like PASCAL, provided a base for the development of machine-independent and somewhat transportable programs. A compiler, which translates the high-level language into a specific machine language, is required for each machine. However, rather than this becoming a complete Tower of Babel, with each machine requiring its own language, we have arrived at a sort of international state, with only half a dozen or so languages required of the user. Each language has its proponents and adversaries. The difficulties involved with writing workable software and writing readable software have not yet been overcome. In the near future, we cannot expect that a single acceptable robot language will come about. However, we can expect the number of languages to narrow down to just a few to facilitate robotic operations by the use of translator programs used for off-line programming, that is, programming not done directly on the robot controller.

Because of the different requirements in speed and complexity for the versatile robot, there are some natural divisions in the requirements for programming languages. As with hierarchical control, we may also think of a hierarchy of programming languages. At the top of the hierarchy, there is a need for a production control language (PCL), such as the one being developed at the National Bureau of Standards, which provides a common interface between the functional levels and applications areas of the automated factory (McLean et al., 1983). There is also a clear need for an off-line programming language, such as a manufacturing language (AML) developed by IBM, or the Stanford artificial intelligence language (AL), which permits the off-line development and simulation of robot control programs that can then be down-loaded to the robot controller for final testing and implementation. Off-line programming permits many programmers to develop programs for a single robot. There is also a need for an on-line programming language for each robot that permits teach control and real-time programming and testing. Furthermore, if we are to use much of the previously developed software and expertise from the artificial intelligence community, then perhaps some attention to symbol-manipulating languages, such as LISP, might be prudent. Rather than answer the unanswerable question about which programming language is superior to all others, we will simply observe that some language is necessary and review some of the languages we might encounter.

First, any computer has a set of instructions, called machine language, that specifies the operation to be performed by the machine as it cycles through its program. For example, on a 68000 microprocessor, the operation code that instructs the machine to add two numbers is the binary number 1101, which is the decimal number 13, or hexadecimal number D. In the binary number system, strings of the binary digits 0 and 1 are used to represent a number. The digits in the string are multiplied by the powers of 2, such as 1, 2, 4, 8, and 16. In the octal number system, strings of octal digits, 0, 1, 2, 3, 4, 5, 6, 7, are used to represent a number. The digits in the string are multiplied by powers of 8, such as 1, 8, 64, and 4096. Octal digits exactly represent groups of three binary digits. In the decimal number system, strings of decimal digits, 0, 1, 2, . . . , 9, are used to represent a number. The digits in the string are multiplied by powers of 10, such as 0, 10, 100, and 1000. In the hexadecimal number system, a number is represented by a string of hexadecimal digits, 0, 1, 2, . . ., 9, A, B, C, . . ., D. Hexadecimal digits exactly represent groups of four binary digits. The digits in the string are multiplied by powers of 16, such as 1, 16, 256, and 65,536, to evaluate the number. A total program can be developed in machine code, as is required on the Heath Company Heathkit HERO I robot. However, it is much easier to use a mnemonic, such as ADD, which can be directly translated into the binary number 1101 but is much easier to remember. An assembly language consists of a set of mnemonics and operations that can be directly translated into machine code. An assembly language program is translated into machine code by a program called an assembler. Assembly language is the most difficult language in which to write a program, because it involves thinking of operations in terms of the basic computer operations, such as register transfers. However, it generally results in the fastest and most powerful way to use a computer.

An interpreter language is a method for implementing a higher level language. Languages of this form have instructions that can be directly interpreted into machine code, often in a one-to-many translation, then executed. BASIC (beginners' all-purpose symbolic instruction code) is one of the most popular interpreters. For example, the instruction X = 1 + 2 may generate a machine code that instructs the computer to add the two numbers and store the result in the memory location labeled X. Interpreters are the easiest languages in which programs may be written, since they can be executed at any time and provide direct feedback about errors. However, it is also the slowest in terms of execution, since the commands are retranslated each time the program is executed.

A compiler program is used with yet higher level languages that interpret commands into many machine language commands and can also optimize the code for such functions as high-speed execution or minimum storage. A compiler requires that a complete program be translated at the same time, so that a single typing error can make it necessary to repeat the compilation process. Another program, called a linker, is often required so that subroutines and other program segments may be compiled separately, then combined to form an executable module.

Each of these approaches uses another program, called an editor, to permit one to enter commands. The editor, assembler, interpreter, compiler, and applications programs are managed by an operating system program, which supervises the use of memory, storage, and machine control. Various operating systems are available for each brand of computer.

We might also want to distinguish between a general-purpose computer and a programmable controller. The main distinction is that a programmable controller is a general-purpose computer used in a limited application without such peripherals as a line printer or a mass-storage device, but with excellent interface signal support. Also, a special logical notation, called ladder logic, is often used in industry to develop programs for programmable controllers. Ladder logic is especially useful for systems with many switches and timing controls.

With the great variety of other languages available, why do we need a robot language? The main reason is that, in controlling a robot, certain primitive operations, such as moving the robot to a position in space and opening or closing a gripper, are used repeatedly, and these operations are specific to robots.

Various levels of robot programming languages have been described by Bonner (Bonner and Shin, 1982). Level 1 is at the microcomputer hardware level and is used for direct robot actuator control. A special machine language called microcode or normal machine language is used at this level. Level 2 consists of a robot path controller high-level language. Teach-by-example programming is indicative of that in which the programmer moves the robot to a specified point in space and pushes a program button to record the actuator locations. A high-level language in the robot controller translates these commands into level 1 commands. Level 3 is also programmed in the robot controller at the primitive motion level, but has basic operations that can be specified by such mnemonics as MOVE. The higher level languages are usually used for off-line programming. Level 4 is a structured programming level in which complex data structures, the use of predefined state variables, and sensor commands are built into the language. Level 5 is a task-oriented level in which English language-type commands may be given. This type of language would depend upon a world model being stored in the computer so that such commands as PICK UP THE TOOL could be interpreted unambiguously. We might even envision a level 6 language in which strategies, motion paths, and a knowledge base are built into the overall system to such an extent that such commands as BUILD TRUCKS might be interpretable.

Some of the characteristics of a good robot language include clarity, simplicity, naturalness, debug and support capabilities, ease of extension, decision-making capabilities, ease of interaction with external devices and sensors, concurrent or parallel operations, and interaction with data base systems.

Currently, the main division of interest in industrial robot applications is between on-line and off-line programming. From the short-term industrial point of view, on-line programming using teach by example offers a simple and low-cost solution that can be quickly mastered by an operator. Also, the microprocessor for on-line programming is simpler and costs less than more complex systems. From the long-term point of view, as CAD and computer-aided manufacturing (CAM) merge together in factories, off-line programming offers advantages of multiuser capabilities, distributed processing, and system integration. Both methods are useful.

Level 1 Robot Languages

Let us now consider some specific robot-programming examples. At the first level, the program would be written in a specific assembly language for high-speed, real-time response. As an example, let us consider a machine language portion of the higher level language called ARMBASIC, which was developed in Z80 assembly language for the Microbot, Inc., MiniMover 5 robot (Microbot, 1980).

The MiniMover 5 uses stepper motors to actuate its 5 joint degrees of freedom and gripper. These motors have four coils, each driven by a power transistor. The command signals are digital. That is, the motors may be driven clockwise or counterclockwise by the execution of the following binary commands on the four drive lines going to each motor.

| Motor drive signals | | | | Hexadecimal value |
d_1	d_2	d_3	d_4	
0	0	0	1	1
0	1	0	1	5
0	1	0	0	4
0	1	1	0	6
0	0	1	0	2
1	0	1	0	A
1	0	0	0	8
1	0	0	1	9

A sequence of commands that goes down the table drives the motor clockwise; a sequence that goes up the table drives the motor counterclockwise. When either end of the table is reached, it is necessary to go to the other end of the table and continue sequentially. The program that performs this control of the motors is called STEP. Only the segment of the program that steps all six motors will be considered. A flowchart of the motor drive program is shown in Figure 4–6. To introduce the level 1 assembly language program, let's first consider the following high-level language description of the program segment.

```
THE MAIN LOOP TO STEP ALL MOTORS
FOR N ::= 1 TO MAX
    FOR I ::= 1 TO 6
        SUM(I) ::= SUM(I) -DELT(I)
        IF SUM(I) < 0 THEN
        SUM(I) ::= SUM(I) + MAX
        PHASE(I) ::= (PHASE(I) + DIRC(I))MOD 8
        SEND MOTOR NO AND PHASE CODE TO OUTPUT
        ELSE DELAY FOR SAME AMOUNT OF TIME
        END IF
    END FOR
DELAY LOOP;COUNT DOWN "CONT"
END FOR
```

The execution of this program is as follows. A form of proportional path motion is implemented in which all motor rotations will end at the same time. Therefore, the value of MAX is the largest number of steps required of any of the six motors. Thus, the first FOR loop will index this largest number of steps. The statements that start and end this loop are

```
FOR N ::= 1 TO MAX
END FOR
```

Since there are six motors, the inner FOR loop will be executed for each motor. The statements that start and end this loop are

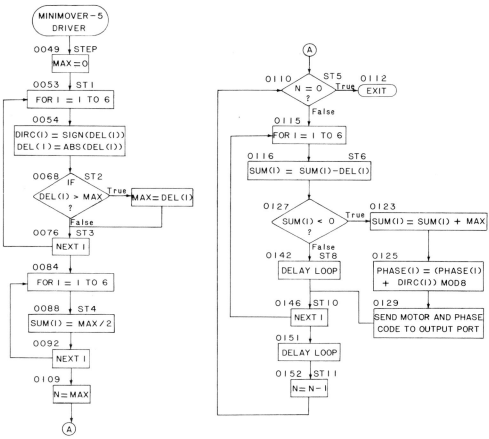

Figure 4-6. Flowchart of the Microbot MiniMover 5 robot, stepper motor drive program. (Courtesy of Microbot, Inc., Mountain View, California.)

```
FOR I ::= 1 TO 6
END FOR
```

All elements of the array SUM have been initialized to the value of MAX/2. The array DELT contains the number of steps required for each motor. The first loop will be performed the maximum number of times. However, only the motor making the maximum number of steps is moved each time. The others are delayed some of the time. If DEL = MAX, then the motor should step each time through the loop. If DEL = MAX/2, then the motor should step each second time through the loop. If DEL = MAX/3, then the motor should step each third time through the loop. Finally, if DEL = MAX/MAX = 1, the motor should only step once. The logic is implemented with the IF-THEN-ELSE statement:

86 | Intelligent Robot Programming

```
IF SUM(I) < 0 THEN
   SUM(I) ::= SUM(I) + MAX
   PHASE(I) ::=(PHASE(I) + DIRC(I))MOD 8
ELSE DELAY FOR SAME AMOUNT OF TIME
```

The algorithm used to determine the number of increments is based on a clever digital differential analyzer used in machine tool control (Seim, 1980). Let's consider the following simplified version for a single motor.

```
DEL = 7
MAX = 10
SUM = MAX/2
   FOR N = 1 TO MAX
     SUM = SUM-DEL
     PRINT "SUM"; SUM
     IF SUM < 0 THEN SUM = SUM + MAX
   NEXT N
```

When this simple BASIC version is executed, the following sequence of values of SUM are generated: –2, 1, –6, –3, 0, –7, –4, –1, 2, –5. Note that exactly seven negative numbers are generated. For each of these, the motor is stepped. This example may be executed with various values of DEL and MAX to verify that the SUM will be less than 0, only DEL times through the loop. Therefore, when this condition is satisfied, the motor should be stepped.

Now, let's consider the assembly language version of this motor drive. For clarity, we will place the high-level command as a comment for the assembly code.

```
LABEL     OP CODE   OPERANDS          ;COMMENTS

          DATA
CONT      DEFS  2
ARRAY     EQU   $
DEL       EQU   $-ARRAY
          DEFS  2 * 6
SUM       EQU   $-ARRAY
          DEFS  2 * 6
DIRC      EQU   $-ARRAY
PHASE     EQU   $-ARRAY + 1
          DEFS  2 * 6
MAX       DEFS  2 * 1
          LD    BC, (MAX)             ;LOAD THE VALUE OF MAX
ST5       LD    A,B                   ;FOR N ::= 1 TO MAX
          OR    C
          RET   Z                     ;THE EXIT CONDITION
          PUSH  BC
          LD    B,6                   ;B IS COUNTER FOR MOTOR
                                      ;LOOP
```

```
              LD    IX,ARRAY              ;REFERENCE POINTER FOR
                                          ;DEL
ST6           LD    H,(IX+SUM+1)          ;FOR I ::= 0 TO 5
              LD    L,(IX+SUM)            ;HL ::= SUM(I) - DEL(I)
              LD    D,(IX+DEL+1)
              LD    E,(IX+DEL)
              XOR   A
              SBC   HL,DE
              JR    NC,ST8                ;IF HL < 0 THEN
              LD    DE,(MAX)              ;HL ::= HL + MAX
              ADD   HL,DE
              LD    A,(IX+PHASE)          ;(PHASE(I) ::=
              ADD   A,(IX+DIRC)           ;  (PHASE(I)+DIRC(I) )
              AND   7                     ;MOD 8
              LD    (IX+PHASE),A
              PUSH  HL                    ;SAVE CONTENTS OF HL
              LD    HL,TABP
              ADD   A,L
              LD    L,A
              JR    NC,ST7
              INC   H
ST7           LD    D,(HL)                ;TABP(PHASE(I) )
              POP   HL                    ;RESTORE HL VALUE
              LD    A,PORT+6              ;PORT FOR MOTOR I
              SUB   B
              LD    C,A
              OUT   (C),A                 ;SEND COMMAND TO MOTOR
              JR    ST10                  ;ELSE
ST8           PUSH  BC
              LD    B,1                   ;DELAY FOR EQUAL TIME
ST9           DJNZ  ST9
              POP   BC                    ;END IF
ST10          LD    (IX+SUM+1),H          ;SUM(I) ::= HL
              LD    (IX+SUM),L
              INC   IX
              INC   IX
              DJNZ  ST6                   ;END FOR
              LD    BC,(CONT)             ;DELAY LOOP
ST11          DEC   BC
              LD    A,B
              OR    C
              JR    NZ,ST11
              POP   BC
              DEC   BC
              JR    ST5                   ;END FOR
;
;             TABLE OF PHASE CODE
;
TABP          DEFB  1,5,4,6,2,0AH,8,9
```

88 | Intelligent Robot Programming

The assembly language version requires one to operate at the basic machine level. This requires a knowledge of the machine architecture as well as the instruction set. The architecture of the Z80 includes two sets of registers; however, only one will be used in this example. The registers are called A, F, B, C, D, E, and H, L. Their contents can be operated in 8-bit byte segments or as 16-bit register pairs for the BC, DE, and HL pairs. The machine also contains other 16-bit registers, such as the index registers IX and IY, the program counter PC, and the stack pointer SP. The A register or main accumulator holds one operand for adds, subtracts, and some other operations; the other operand may come from the other registers or memory. This permits two operand instructions. The F register holds the conditional bits or flags, such as the negative, zero, and carry bits, which may be set as the result of an operation. Let's go through the assembly language program statement by statement, just as it is executed.

The first statement simply loads the contents of the memory location labeled MAX into the BC register pair. Note that the variable names, such as MAX in the previous BASIC programs are implemented by assigning a memory location to the variable name and storing the value of the contents in the memory location. For example, MAX=7 may correspond to assigning memory location 1000 as MAX and storing 7 in the memory location. The instruction LD BC,(MAX) puts the contents of (MAX) into the B register and the contents of (MAX+1) into the C register. Note that the syntax first gives the operation LD, then the operands BC and (MAX). Since BC is a register, its address is known to the machine and thus its contents are simply referred to as BC; however, for memory location MAX, the address of this location, for example, 1000, would be obtained by using the syntax MAX. To obtain the contents of the location, we must use the syntax (MAX).

The next statement starts the loop for the number of steps:

```
LD A,B
```

Suppose the value of (MAX) is 7. In the previous instruction, this value was transferred to the B register. The value 7 is now transferred to the A register for ease of further operations. If the value (MAX) were less than 255 and (MAX+1) had previously been initialized to a value 0, the C register should contain a zero.

The next instruction simply sets the flags to determine if the contents of the A register equal 0. This is accomplished by

```
OR C
```

This instruction logically ORs the contents of the A and C registers and sets condition flags in the F register. Since C contains 0, the result of the OR instruction is simply equal to the contents of A. Only when all the bits of A are zero will the zero flag in the F register be set.

The next instruction

```
RET Z
```

returns to the instruction immediately following the one that called this program if the zero flag Z is set. The loop from 1 to MAX is therefore implemented in an A register, then later decrementing A and returning when its value is zero. Note that this comparison at the beginning of the operation counts the correct number of iterations. For example, if MAX=7, the sequence would be 7, test, 6, test, 5, test, 4, test, 3, test, 2, test, 1, test, 0, exit.

The next few instructions set up the loop counter for the six motors by first saving the contents of BC and then putting the value 6 into the B register, that is,

```
LD   B,6
```

Next, the memory address of the array ARRAY is loaded into the index register IX. This technique permits IX to be used as a pointer to values of the array, using the indexed addressing mode. Two arrays are used. One is used to store the six values of SUM(I) and the other to store the values of DEL(I). These values are stored at locations IX+SUM and IX+DEL, with two memory words used for each parameter. For example, SUM could equal 0 and DEL could equal 12.

The next four statements put particular values of SUM(I) into the HL register and DELT(I) into the DE register. Note that the 16-bit integers are stored with the most significant byte in the low-order memory location and the least significant byte in the high-order memory. Thus, the individual bytes are transferred to the four individual registers in the following manner.

```
LD   H, (IX + SUM+1)
LD   L, (IX + SUM)
LD   D, (IX + DEL+1)
LD   E, (IX + DEL)
```

Note the difference between loading the HL register by LD HL, (IX + SUM).

The next instruction,

```
XOR  A
```

performs a logical, exclusive OR of the contents of the A register with itself and sets flags. The exclusive OR of two 1 bits is a zero, so this instruction essentially forms a zero result and resets the carry and sign flags and sets the zero flag.

The next instruction,

```
SBC  HL,DE
```

subtracts the contents of DE from HL and puts the result in HL. The next instruction,

```
JR   NC, ST8
```

90 | Intelligent Robot Programming

jumps or branches to location ST8 if the results of the subtraction SUM(I) did not set the carry flag, that is, were positive.

If the result SUM(I) was negative, then the next two instructions add back the value of (MAX), that is,

```
LD DE, (MAX)
ADD HL, DE
```

The next instruction loads the PHASE(I) value into the A register, that is,

```
LD , (IX+PHASE)
```

The phase value is stored following the DEL(I) values.

The next instruction adds the direction value DIRC(I) and saves the result in the A register, that is,

```
ADD A , (IX+DIRC)
```

To compute the sum modulo 8, we simply throw away any bits past the third bit or numbers greater than 7. This is accomplished by the logical AND of the contents of the A register and the number 7, that is,

```
AND 7
```

This result is now stored as the new value of PHASE:

```
LD (IX+PHASE) , A
```

At this point, the HL register contains the updated value of SUM(I). It is now saved for the output to the motor and the next iteration computation by pushing it on the stack, that is,

```
PUSH HL
```

The next sequence of instructions sets the phase codes to control the direction and step code for the motor. The address of the stepper motor codes is first loaded into the HL register, that is,

```
LD HL , TABP
```

Next, the value of the phase offset is added to the address:

```
ADD A,L
```

This value is now placed back into the L register so that the HL pair now points to the appropriate phase value in the table:

LD L,A

A jump to ST7 is now made if the ADD instruction did not result in a carry. If a carry did result, the H register is incremented, that is,

INC H

At ST7, the command values are output to the motor. First, the motor code is loaded into the D register by

ST7   LD D, (HL)

The old value of SUM(I) is put back into the HL register by

POP HL

The port number is now loaded into the A register by

LD A , PORT+6

The loop counter value is subtracted from the A register

SUB B

The resulting port value is now placed in the C register by

LD C,A

Finally, the command is sent to the motor by

OUT   (C),A

The program now jumps to ST10. At this location, the value of SUM(I) is updated in the memory array by

ST10   LD (IX+SUM+1), H
                            LD (IX+SUM), L

Also, the array counter is incremented by 2 to skip past the previous value by

INC   IX
INC   IX

The B register contains the loop counter. The next instruction decrements this loop counter and branches to ST6 to start the next iteration if the result is not zero by

```
DJNZ ST6
```

If the result is zero, the loop is completed and the next step sequence can be started. To control the motor speed a delay loop between the steps is started. This permits a speed control of the movements. A counter value is loaded into the BC register by

```
LD    BC,(CONT)
```

The BC value is then decremented by

```
ST11 DEC BC
```

The most significant byte is then loaded into the A register by

```
LD    A,B
```

The logical OR of this and the C register is then computed by

```
OR    C
```

When both bytes of BC contain zero, this operation will set the zero flag. The next instruction tests this condition:

```
JR    NZ,ST11
```

After the delay, the saved value of BC, which is (MAX), is started again by

```
POP   BC
DEC   DC
JR    ST5
```

One might ask why it is necessary to write the motor drive program in assembly language, since it requires a great deal of effort to read and write such a machine-dependent language. The main answer is speed. Some of the motor drive commands might operate at rates of a million per second. High-level languages have not yet reached this operating speed. Assembly language and microprogramming, which we have not discussed, are both much easier than building an electronic circuit to perform the operation. However, from the user's viewpoint, higher level languages are definitely desirable. Let's now look at some of the higher level robot languages.

The second level of programming uses a language designed for the primitive operations of the robot and is operable on the robot controller. For the MiniMover, a language called ARMBASIC is used, which has the following primitives.

@STEP   Moves the manipulator with the delay between pulses to the motors and the number of steps for each motor as a parameter.
@CLOSE   Closes the gripper with the delay for the gripper motor as a parameter.
@SET   Allows the manipulator to move under manual control with delay as a parameter.
@RESET   Zeroes the arm position and motor currents and is used for arm position initialization.
@READ   Returns the current position of the manipulator by returning the values of the six position registers for the six drive motors.
@ARM   Selects the port number ARMBASIC uses. This permits the control of the robot arms.

A simple example of the type of program that can be easily written in ARMBASIC is a pick-and-place program. Suppose an object is to be picked up at location A and placed at location B on a flat surface. If the object is to be picked up from the top, we may need to insert intermediate positions, such as A' directly above A and B' directly above B, so that slower approach and depart trajectories from A to A' and B' to B may be used. Assuming that the robot starts at A', the desired sequence of operations is as follows.

1. Move down to A.
2. Close gripper slowly.
3. Move back to A' at slow speed.
4. Move to B' at high speed.
5. Move to B at slow speed.
6. Open gripper slowly.
7. Move to B' at slow speed.
8. Move to A' at high speed.

For performance of these movements, we must also specify the locations of the STEP primitive and the speeds for the delays. The position vectors are needed in terms of the number of steps required for each motor. This number of steps is directly related to the angle of motion required and may be called joint-step coordinates. Suppose the number of joint steps required to move from A' to A, A' to B', and B' to B are P, Q, and R, respectively, where

$P = (P1,P2,P3,P4,P5,P6)$
$Q = (Q1,Q2,Q3,Q4,Q5,Q6)$
$R = (R1,R2,R3,R4,R5,R6)$

Let the slow speed be specified by a delay S, the high speed by a delay H, and the gripping speed by a delay G. Finally, let the number of steps required to open the gripper be CG. The following ARMBASIC program would implement this pick-and-place task repeatedly.

```
10      STEP        S,  -P1,  -P2,  -P3,  -P4,  -P5,  -P6
20      CLOSE       G
30      STEP        S,   P1,   P2,   P3,   P4,   P5,   P6
40      STEP        H,   Q1,   Q2,   Q3,   Q4,   Q5,   Q6
50      STEP        S,   R1,   R2,   R3,   R4,   R5,   R6
60      STEP        G,    0,    0,    0,    0,    0,   CG
70      STEP        S,  -R1,  -R2,  -R3,  -R4,  -R5,  -R6
80      STEP        H,  -Q1,  -Q2,  -Q3,  -Q4,  -Q5,  -Q6
90      GOTO        10
```

Upon executing this program, you must be ready to keep placing objects at position A. The manufacturer also provides numerous example programs to illustrate the use of the MiniMover robot. Other examples are described by Hemenway (1983).

This example has one basic problem from the user's viewpoint. For each location, the number of motor steps in the joint-step coordinates must be known. This requires some rather tedious calculations, especially for a large number of positions. Two solutions are offered for this problem.

The first is called teach-by-example programming. With this method, a human programmer interactively moves the robot to a desired position and pushes a button to record the joint angles or position. This procedure can be repeated for each desired position. The human programmer must also specify velocities and tool operations; however, the difficult calculation of joint angles or joint-step values can be done by the computer.

The second approach is to develop the transformations from a more natural coordinate system, such as Cartesian coordinates to the joint angle coordinates. Note that both the position (x,y,z) as well as the orientation angles of the gripper or tool are required to determine the joint angles. For the Microbot MiniMover, there are five joint angles. In general, six are required. Given the joint angles of the robot, it is relatively easy to determine the forward, kinematic problem solution. However, given the Cartesian coordinates, it is relatively difficult to determine the corresponding joint angles. This is called the inverse kinematic problem and solution. The inverse problem is of course solved for each robot in use.

The difference between teach-by-example programming and the coordinate transformation method is more fundamentally in the operation than in the mathematics. Let's now explore teach-by-example programming to illustrate the power of this approach.

Teach-by-Example Programming

Let's use the Cincinnati Milacron heavy-duty T3 (HT3) hydraulic-powered robot language for an example of a higher level language (Operating Manual for the Cincinnati Milacron T3 Industrial Robot). This language provides for teaching by example (level 2), as well as transformations, primitive motions, and external control, which are level 3 features (Holt, 1977). Certain conditions must be met to actually operate the robot. The power must be turned on for the controller and the hydraulics. Then, the operator may enter either a teach mode or an automatic mode. Safety is always the primary consideration. Although there are manual stops that restrict the robot's motions somewhat, the robot, like any moving machinery, requires careful usage. In the teach mode, a teach pendant, such as that shown in Figure 4–7, permits the operator to position the robot, program a point, operate the grippers, and single step through a sequence of programmed points. The teach pendant has a large button for emergency stop. Depressing this button disables the hydraulic power, which stops the powered motion; however, the arm can still drift downward. The position buttons permit the operator to position the arm in the coordinate systems selected. Rectangular, cylindrical, or hand coordinates, as shown in Figure 4–8, may be selected. Note that the robot controller assists the operator by permitting the selection of different coordinate systems. The teach pendant also has three sets of buttons to control orientation in terms of yaw, pitch, and roll. Tool functions may also be selected for two possible tools. When these buttons are depressed, the tool status changes. If the tool is open, it closes, and vice versa. When the robot is positioned at the desired location, the program button is pressed, which stores the actuator locations in memory. Such functions as velocity may also be selected by depressing the function

Figure 4–7. Teach pendant for the Cincinnati Milacron T3 robot. (Courtesy of Cincinnati Milacron.)

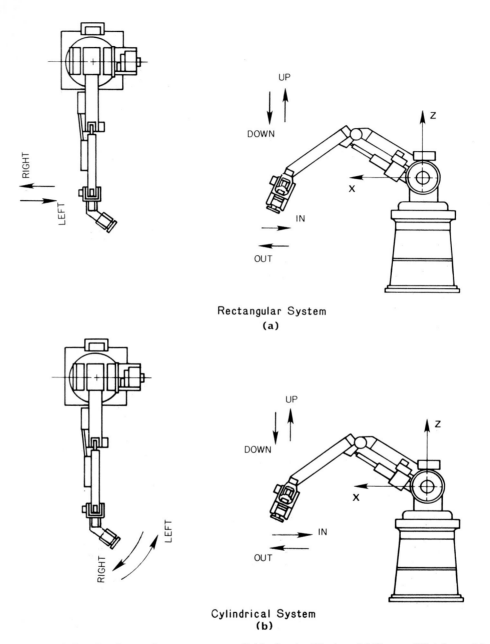

Figure 4–8. Teach coordinate systems available for the Cincinnati Milacron T3 robots. (a) Rectangular coordinates. (b) Cylindrical coordinates. (c) Hand coordinates. (Courtesy of Cincinnati Milacron.)

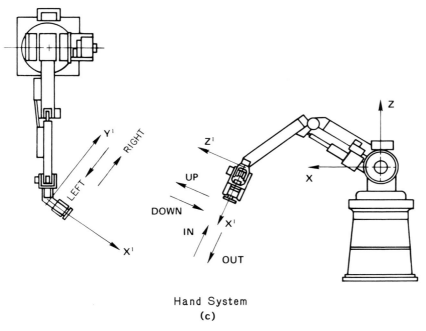

Hand System
(c)

Figure 4–8 (continued)

button and entering the desired value. A cathode-ray tube (CRT) and keyboard are also available to monitor the operation and permit program entry or modification or activate automatic mode.

A program must start with the robot in the home position. The home position provides a common reference position on all the manipulator axes. A program is developed by using the teach pendant to move the robot to the desired position. Then a function, velocity, and tool dimension are selected. If no particular function is desired, a no operation NOP may be selected. A program button is then depressed to store the information in the robot's memory. A velocity, which may vary from 0 to the maximum velocity, may be selected. The tool dimension specifies a point a given distance in front center of the wrist faceplate, which provides an orientation reference. A program consists of a main line sequence, and up to 63 non-main line sequences. A maximum of 999 points is permitted in a sequence. A main line sequence is shown in Figure 4–9. Note that the home position is numbered as position 00.001. The first two digits refer to the sequence number, which is 00 for the main line sequence. The other three digits refer to the point in the sequence. After a sequence has been programmed, a close path operation is performed to a point in the main line sequence that has a NOP function assigned to it. This point will become the cycle start point. Also, the cycle start position is numbered 00.002. The cycle start must contain a NOP function.

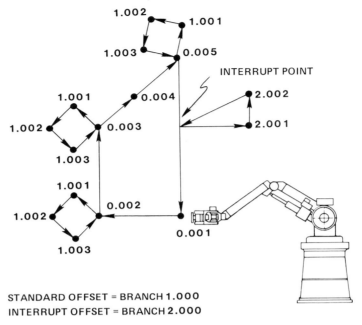

Figure 4–9. Diagram of the Cincinnati Milacron T3 program, showing main line and subroutine sequences. (Courtesy of Cincinnati Milacron.)

Let's now consider a more realistic example that uses signals. We will look at a diagram of the process shown in Figure 4–9. A program corresponding to this diagram is shown below.

```
SEQ. POINT    LOCATION             FUNCTION
0.001         1 or Home            NOP
0.002         2                    PERFORM,1,C ALL
0.003         3                    PERFORM,1,C ALL
0.004         4                    NOP
(Cycle start)
0.005         5                    PERFORM,1,C ALL
(Close path, 0.001)

(PERFORM 1)
1.001                              NOP
1.002                              NOP
1.003                              NOP
(PERFORM 2 on interrupt, close path subroutine, relocatable)
2.001                              NOP
2.002                              NOP
(Abort routine for main line program)
```

As illustrated, the main line program starts at statement 0.001, which is the home position. It then moves to position 2 and performs the subroutine sequence 1. It then moves to position 3 and again performs subroutine 1. It then moves to position 4 and does nothing. It then moves to position 5 and again performs subroutine 1. The interrupt subroutine 2 could be an abort routine, which is performed if something goes wrong during the program execution.

This simplified routine shows the main line and subroutine structure but is not directly executable since the velocity, tool dimension, and tool status are not given.

Let's now consider a more realistic example that has all the parameters specified and also uses input and output signals. The robot must pick up a part from the incoming conveyor, load it into the machine, and then unload it and place it on the outgoing conveyor. Signal 1 will indicate to the controller that a part is ready to be picked up. Signal 2 will indicate that the machine is ready. These are both input signals to the robot. The robot will send out signal 3 when the part is loaded into the machine so that it can start its operation. The program is shown below.

SEQ. POINT	LOCATION	FUNCTION	VEL.	DIM.	STATUS
00.001	HOME	HOME	1	1	
00.002	1	NOP	20	1	
00.003	1	PERFORM,1,+S01 (CLOSE PATH,00.002)	0	1	
01.001	1	TOOL,1	0	1	+N
01.002	2	TOOL,1	10	1	-N
01.003	4	NOP,CONTINUE	20	1	
01.004	3	OUTPUT,+01	5	1	
01.005	3	DELAY,5.0	0	1	
01.006	3	TOOL,1	0	1	+N
01.007	4	WAIT,U,+S02	20	1	
01.008	3	TOOL,1	5	1	-N
01.009	3	OUTPUT,+03	0	1	
01.010	4	NOP,CONTINUE	20	1	
01.011	5	TOOL,1 (CLOSE PATH)SUBROUTINE	10	1	+N

The program has a main sequence consisting of the first three statements and a subsequence consisting of the next eleven statements. At each statement, a function, velocity, and tool function may be specified. Let's examine the program in detail.

The first statement, 00.001, corresponds to the HOME position of the robot. The next statement sets the velocity at 1 inch/sec and the tool dimension at 1 inch. The tool dimension also determines a position with respect to the wrist faceplate in which straight-line controlled path motion takes place. That is, the points, which are the control points for interpolated motion, are specified by the tool dimension. The next statement instructs the robot to wait for signal 1 from the input conveyor before moving to sequence 1. Upon receiving signal 1, the robot performs sequence 1, which starts at statement 01.001. At the start of the sequence the robot opens tool 1 but stays at the same position. At step 2, the robot moves to position 2 and closes the gripper on the part. At step 3, the robot moves in front of the machine. At step 4, the robot slows down and presents the part to the machine. Next, at step 5, the robot waits for the machine to grasp the part. At step 6, the gripper opens. At step 7, the robot waits, untimed, for the machine to finish. Next, the robot opens the gripper, enters the machine, and regrasps the part. At step 9, the robot signals the machine to release the part. Next, at step 10, the robot backs out of the machine. Finally, at step 11, the robot moves to the output conveyor and releases the part. The close path to position 2 returns control to statement 2 in the main line sequence, where the robot again waits for a part to be present. The sequence is then repeated continually.

This simple example illustrates the unique requirements of a robot programming language for a machine that can interact with other machines through input and output signals, open and close tools, and move to different points at different speeds. The T3 language also has provisions for external control from other computers, which permits off-line programming.

As another example, consider the following VAL program for a Unimate/Westinghouse robot (User's Manual to VAL). This sample program performs the following tasks. The robot first waits for a part to be in place in a feeder. It then picks up the part and carries it to an inspection station. At this position, it sends a signal to the inspection station that a part is in place. The station then determines whether the part is type A or type B. If the part is type A, one subroutine is performed. If the part is type B, another subroutine is performed. If the part is neither type A or B, a process reject is performed. The cycle repeats indefinitely.

The program is also set up to branch to an emergency subroutine if this condition is indicated by input signal 7, at any time from the start of the program until the IGNORE statement.

```
       REMARK  Start of program
       REMARK
       REMARK  Initialize signal line
       SIGNAL  -2
       REMARK  Make sure hand is initially open
       OPENI   100.00
       REMARK
10     REMARK  Start of loop to process parts
       REMARK
       REMARK  Start looking at emergency signal on input
               channel 7
       REACTI  7,EMERGENCY ALWAYS
       REMARK  Wait for "part in place" signal on input
               channel 1
       WAIT 1
       REMARK  Pick up part from feeder
       SPEED 200.00
```

```
        APPRO PART, 50.00
        MOVES PART
        CLOSEI 0.00
        DEPARTS 50.00
        REMARK Move to inspection station
        APPRO TEST, 75.00
        MOVES TEST
        REMARK Stop checking for emergency signal
        IGNORE 7 ALWAYS
        REMARK Signal that part is in place
        SIGNAL 2
        REMARK Wait for "inspection done" signal on input
               channel 6
        WAIT 6
        REMARK Withdraw from inspection station
        DEPART 100
        REMARK Reset "part in place" signal
        SIGNAL -2
        REMARK Test results of inspection; first for part
               "A"
        IFSIG -3,4,-5, THEN 20
        REMARK Then for part B
        IFSIG 3,-4,-5, THEN 30
        REMARK Part is neither "A" nor "B"--process
               reject
        GOSUB REJECT
        GOTO 40
20      REMARK Process part "A"
        GOSUB PART,A
        GOTO 40
30      REMARK Process part "B"
        GOSUB PART,B
40      REMARK Part processing completed, get another part
        GOTO 10
        REMARK End of program
```

This language appears similar to BASIC because of the REMARK, GOTO, and GOSUB statements. However, the robot primitives are embedded in the language. Let's also consider the operation of this program.

The first executable statement, SIGNAL, transmits a negative signal on channel 2. The next statement, OPENI, immediately opens the gripper 100 millimeters. The REACTI statement initiates continuous monitoring of signal 7. The next statement, WAIT, puts the program in a wait state until a signal is received on channel 1. The next statement, SPEED, requests that the next arm motion be performed at a speed twice as great as normal, which is speed 100 millimeters/second. The APPRO statement moves the end effector to the position specified by array PART, offset by a distance of 50 millimeters in the z direction. The CLOSEI positions the tool near the part. Next, a

MOVES command moves the tool to the position and orientation specified by the array PART. The CLOSEI closes the gripper. The DEPARTS statement moves the hand a distance of 50 millimeters along the current axis of rotation of the last joint. Since the distance is positive, the hand retracts. Next, the APPRO statement moves the hand to the position specified in array TEST, offset by a distance of 75 millimeters. The MOVES statement then moves the hand to the position and orientation specified in array TEST. The IGNORE statement stops testing for the emergency signal. A positive signal is then sent out on channel 2 to inform the inspection unit that a part is in place. A WAIT of 6 seconds is then initiated. The DEPART statement moves the hand back a distance of 100 millimeters. A negative signal is then sent out by the SIGNAL statement. A test is now made. The IFSIG command tests for a negative signal on channels 3 and 5 and a positive signal on channel 4. If these exact conditions are met, program control branches to statement 20. A similar test with different conditions is made and can send control to statement 30. Otherwise, the GOSUB statement branches to REJECT. At statements 20 and 30, various processing on parts A and B can be accomplished by the GOSUB statements. Also, even though the positions, such as PART and TEST, are arrays, the values in these arrays must be specified by teach programming or some other method.

Now let's look, as an example of a level 4 language, at an off-line program example using the AML language developed by IBM (Taylor, et al., 1982). Since AML is an off-line programing language, all positions and orientations must be specified in the program. Upon execution, a calibration procedure must be accomplished to ensure that program positions and physical positions correspond. The sample program moves to a position above a part, checks to see if the part is present with a sensor, grasps the part if it is present, and moves back. Since AML is structured, all variables are specified at the beginning of the program.

```
;GLOBAL DATA DEFINITION
PICKUP
   POINT=(10.000,15.000,10.000,90.00,180.00,90.00)
DISTANCE = 7.0
FUNCTION GETPART
GLOBAL PICKUP POINT, DISTANCE
BEGIN
        SPEED = FAST
;APPROACH SOME DISTANCE ABOVE THE POINT TO START
APPROACH DISTANCE FROM PICKUP POINT
;MOVE TO THE POINT--STOP AS SOON AS WE CAN SEE IT
   WITH SENSOR
MOVE PICKUP POINT UNLESS PART PRESENT = = ON
;CHECK IF IT IS PRESENT
IF PART PRESENT = = ON THEN
        BEGIN
;               PART WAS PRESENT. MOVE TO THE POINT
                WITH A SLOW SPEED
MOVE PICKUP POINT WITH SPEEDSCHED(1)
```

```
;GRASP THE PART
CLOSE
        END
ELSE
        BEGIN
;       PART WAS NOT PRESENT, PRINT ERROR MESSAGE
        WRITE(ERROR - PART WAS NOT PRESENT IN
          FIXTURE)
        END
;MOVE UP A SAFE DISTANCE FROM CURRENT POSITION
DEPART DISTANCE
END
```

The English-like nature of AML makes the program easy to read; therefore, it will not be described in detail.

Finally, let's consider another level 4, off-line programming example using the language AL developed at the Stanford Artificial Intelligence Laboratory (Mujtaba and Goldman, 1979). The following program segment was written to grasp the handle of a pencil sharpener and place it into an assembly fixture. The AL program statements are separated by ";" and surrounded by the reserved words BEGIN and END.

```
BEGIN
DEFINE ARM = "BARM"
DEFINE STOP ARM = "BARM"
DEFINE FING = "BHAND"
DEFINE HAND = "BHAND"
DEFINE PARK = "BPARK"
DEFINE INCH = "2.54*CM"
DEFINE INCHES = "INCH"
DEFINE OZ = "28*GM"
DEFINE / = "COMMENT"
DEFINE $ = "COMMENT"
DEFINE DIRECTLY = "WITH ARRIVAL=NILDEPROACH WITH DEPARTURE=NILDEPROACH";
DEFINE NO ARRIVAL = "WITH ARRIVAL = NILDEPROACH ";
DEFINE NO DEPARTURE="WITH DEPARTURE=NILDEPROACH ";
DISTANCE FRAME ORIGIN;
        ORIGIN =FRAME(NILROTN,VECTOR(0.0,40,.13) );
FRAME JIG BOTTOM;
        JIG BOTTOM = ORIGIN*TRANS(NILROTN,
                        VECTOR(14.8,3.06,1.06));
DISTANCE VECTOR UP3;
        UP3 = 3*ZHAT;
(NOTE "BEGINNING");
/ ASSEMBLE HANDLE;
BEGIN
/ INITIALIZE POSITION OF ARM;
BARM == BPARK
MOVE BARM TO BPARK;
DISTANCE FRAME HANDLE REF, HANDLE GRASP;
        HANDLE REF = ORIGIN*TRANS(NILROTN,
          VECTOR(2.5,8,1.06) );
HANDLE GRASP = HANDLE REF*TRANS(YHAT
                ROT-180,VECTOR(0.8,0,0));
AFFIX HANDLE GRASP TO HANDLE REF RIGIDLY;
```

```
OPEN HAND TO 0.6;
MOVE ARM TO HANDLE GRASP;
CENTER ARM;
WHILE HAND < 0.2
     DO BEGIN
            OPEN HAND TO 0.6;
            MOVE ARM TO .+ ZHAT*3;
            ABORT("I THINK HANDLE IS NOT IN THE
               RIGHT POSITION: PLEASE RECTIFY");
            MOVE ARM TO HANDLE GRASP DIRECTLY
            CENTER ARM;
       END;              /HALLELUJAH, WE GOT THE
                              HANDLE;);
AFFIX HANDLE TO GRASP TO ARM RIGIDLY;
MOVE HANDLE REF TO JIG BOTOM*TRANS(ZHAT ROT
  -86,ZHAT*0.5);
/NOW DROP THE HANDLE;
/ OPEN HAND TO 0.5;
UNFIX HANDLE GRASP FROM ARM;
MOVE ARM TO .+ UP3;
CLOSE HAND TO 0.0;
/NOW SIT ON THE HANDLE;
MOVE ARM TO .-UP3
       ON FORCE(ZHAT) > 12 DO STOP ARM;
ARM == HANDLE GRASP+0.3*ZHAT;
MOVE ARM TO .+0.2*YHAT
/ARM HAS PUT HANDLE FLUSH ON THE FIXTURE
$ WE CAN NOW DESTROY ALL AFFIXMENTS TO HANDLE
   SINCE WE DON'T NEED IT ANY MORE;
END;
```

These examples illustrate the various types and levels of programs used in robot programming. The need for an industry standard is clear, but one has not been achieved. Fortunately, the basis of a new language may be mastered in a relatively short time, especially if one already has a solid background in programming.

Off-Line Programming

Off-line programming is the process of programming the robot on an off-line computer. An illustration of this process using a McDonnell Douglas Automation Company, McAuto robotics simulation system is shown in Figure 4–10. This concept is important, since it ties together the concepts of computer-aided design with the concept of computer-aided manufacturing. The design process and the manufacturing processes can be integrated, simulated, tested, and verified before actually setting up the manufacturing process. To accomplish this integration will require advances in all the areas of intelligent robot programming discussed in this chapter.

Questions

1. Compare the features of the on-line programming languages, T3 and VAL, with the off-line languages, AML and AL. Describe the features present in each class and the advantages and disadvantages of each.

Figure 4–10. An off-line programming example. (Courtesy of Cincinnati Milacron.)

2. Compare the features of interpretative and compiled robot programming languages. List the advantages and disadvantages of each.

3. Describe how learning by experience may be implemented in an intelligent robot system (*hint:* pattern recognition and tree searches).

4. Discuss the direct application of optimal linear control theory for robot control.

5. Discuss the problems in robot control for the four situations:

   a. Robot path not specified and no obstacles
   b. Robot path specified and no obstacles
   c. Robot path specified but obstacles in path
   d. Robot path not specified but obstacles in path

Which problem is most commonly encountered by humans; by industrial robots? Which problem is most difficult?

Robot Sensors

> Robots are far from being at the end of their evolutionary tether. In the near future they will be made adaptable to a greater variety of manual tasks. The trick is to give robots at least one human sense, rudimentary eyesight. An otherwise puerile job is baffling to a robot if parts are not oriented at the pickup point.
>
> Joseph F. Engelberger (1980)

Almost all industrial robots use internal sensors, such as shaft encoders to measure rotary joint position and tachometers, which measure velocity to control their motions. Most controllers also provide interface capabilities so that signals from conveyors, machine tools, and the robot itself may be used to accomplish a task. However, external sensors, such as visual sensors, can provide a much greater degree of adaptability for higher level robot control as well as add automatic inspection capabilities to the industrial robot. Visual and other sensors are now used in such fundamental operations as material processing with immediate inspection, material-handling with adaptation, arc welding, and complex assembly tasks. A new industry of robot vision companies has emerged. This chapter provides an introduction to this new, important technology.

To understand what sensors might be desired or needed for robots to optimally perform tasks, it is useful to first review the human senses to see how they are used as we live and work. In so doing, we can appreciate the complexity of the human and perceive some of the problems encountered in equipping robots with simulated senses. We are all familiar with five of the seven senses of the human: sight, smell, taste, hearing, and touch. These give us an incredible amount of information about our environment that enables us to make decisions about how to adapt to new or unexpected situations.

Let's start with sight. One of the most important things sight allows us to do is to select proper, safe paths for motion. Binocular vision and other perceptual cues, such as object occlusion, permit us to judge the distance of objects. Color vision permits rapid discrimination of millions of different shades of light and color. The spatial acuity provided by high-density rod and cone receptors allows us to perceive a minute speck of dust in optimum lighting. Our brightness sensitivity is so acute that we can see the light of a single candle at a distance of 30 miles on a very dark night. Automatic brightness control permits rapid adjustment from very light to very dark environments. We depend on our eyes to give us most of the information we need. It is estimated that 70 percent of the information that reaches the brain comes through our visual sense. An interesting experiment (Buffington, 1983) illustrates the dominance of vision in our understanding of the world about us. It is also easy to perform. Close your eyes, and trace the number 2 with your finger on your forehead. Does it feel reversed? Now trace the same number on the back of your head. Does it feel normal? This experiment shows that, just because our eyes face forward and dominate our senses, other senses, such as tactility, are important for correctly performing tasks.

Hearing is, like vision, stereoscopic, and permits us to judge the direction and distance of a sound. This sense is well developed even before birth and works best when we are asleep. It is so acute that many mothers can hear the breathing of a newborn infant in another room. We also use hearing to select proper forms of motion, especially when visual cues are missing or obstructed, as when we hear a car coming before we see it. We also use our hearing in making decisions. For example, many very experienced automobile mechanics can simply listen to an engine running and correctly identify any problems. We are able to distinguish many different tones and wavelengths, which enables us to distinguish and identify millions of objects and phenomena in our world.

Smell is a chemical sense, olfaction, as is taste, gustation. Our olfactory senses enable us to distinguish many objects and phenomena without the use of any other senses. For example, we can usually tell what foods are edible or ripe by smelling them, even though our visual cues may indicate otherwise. Olfaction is particularly important in enabling us to identify invisible or hidden elements, such as gases. Taste is also important in determining the potability of food. The four taste qualities—bitter, sour, salty, and sweet—help provide us with the impetus to obtain essential nutrients. Many people would say their gustatory senses are too well developed. However, we use our taste sensors to again discriminate and distinguish many objects. For instance, we can detect the presence of minute amounts of metal in our foods (which is why you don't use metal utensils to cook foods with delicate flavors), as well as the presence of some gases and chemicals that are neither seen nor smelled.

Touch includes more sensitivity that we often think about. Sensors for pressure, temperature, and pain are embedded in our skin by the thousands. For example, there are about 3,000,000 pain sensors, 500,000 pressure sensors, and 200,000 temperature sensors distributed unevenly throughout the human body, mainly on the surface. For example, there are 232 pain sensors per square centimeter behind the knee, 60 per square centimeter on the thumb pad, and 44 per square centimeter on the tip of the nose. Again, we can use this sense to help us identify and distinguish objects and phenomena when our other senses fail or are obstructed. For instance, we can feel and identify a caterpillar crawling on our backs and take appropriate action without using our other senses.

These are the five commonly known senses, but we have two others that are also very important and of which we are not usually consciously aware. Some sense receptors located in the tendons, joints, and muscles inform the brain of the position and movements of the entire body. This kinesthetic sense permits us to walk without watching our legs, to tense a muscle without looking at it, or to touch our finger to our nose with our eyes shut. The ability to walk uses this kinesthetic sense, but it also requires a sense of balance. This balance is provided by the vestibular sense, located mainly in the inner ear, informing us whether we are upside down or right side up, speeding up or slowing down, or rising or falling. It is the manipulation of this sense that gives us such a thrill on a roller coaster ride.

There may be other senses humans possess, such as a magnetic sense, but these are still in some dispute. These seven certainly seem sufficient to allow us to adapt to our environment and perform our work. However, it is characteristic of humans to want to improve on even these extremely versatile senses with inventions that extend or amplify them. These extensions include microscopes, telescopes, sonar, radar, x-ray, and infrared devices or detectors, and Geiger counters. They give us far more information about our world than we could receive using our unaided senses. Equipping intelligent robots with sensors includes considering not only our own senses, but these extensions, as well, in enabling them to perform certain tasks. Sensing techniques, such as the ability to see in all directions at once, could be built into a machine. Parents often seem to have eyes in the back of their heads. The robot actually could. Similarly, sensors that measure force, torque, proximity, temperature, and magnetic or electric field properties could be implemented to measure these quantities at isolated points or at several neighboring points to permit recognition of an object. For the universal machine, all these senses may be needed.

5.1 Robot Sensor Classification

Sensors may first be classified as internal or external, in which the external sensors, such as vision or tactility, are not included as inherent components of the robot controller, but the internal sensors, such as shaft encoders, are built into the manipulator. This classification is indicative of the early stage of the evolution of robots. Eventually, it is likely that all useful sensors will be integrated into the robot design.

Another classification is based upon the function performed by the sensor. Although many types of sensors for robots are still being invented, there are some generic groupings of the different types, as shown in Figure 5–1. For the acquisition stage, in which information is being collected and the robot manipulator is not in contact with the part, a noncontact sensor must be used. This assumes that the only sensors are those on the robot. External sensors, such as a separate tactile table, could be used to sense shapes. Of the various types of noncontact sensors, there is a natural division into those that measure a point response and others that give a spatial array or measurements at neighboring points of information. An example of a point-measuring device may be found in the distance-measuring ultrasound devices developed by Polaroid. These devices measure the distance to the nearest object within a cone of information-

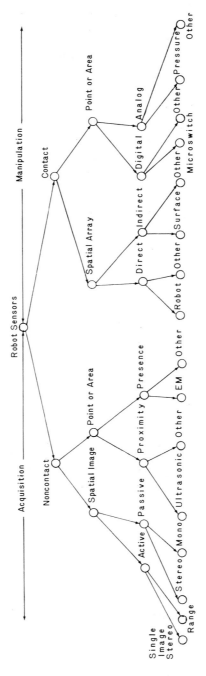

Figure 5-1. Classification of sensors available for use on an intelligent robot.

109

collection space. A camera is the most common example of a device that measures spatial information. Point or area sensors may be further divided into those that measure proximity and those that measure the presence of an object. Further subdivision of sensors into devices that measure spectral range, such as infrared, visible, and x-ray, may be developed.

A similar grouping may be developed for contact sensors, which are most useful in the manipulation stage of a process. Contact sensors may simply measure touch or may measure force or torque. The most common touch sensor is a simple switch that closes when it touches a part. Force or torque sensors may work according to Newton's law, that force is equal to the product of mass times acceleration, or torque is equal to the product of inertia times angular acceleration. A simple force sensor might measure acceleration with an accelerometer to sense force. These sensor categories may also be divided into whether a direct or indirect method of measurement is used. For example, the force could be measured directly at the robot hand or indirectly by its effect on the working surface. Force and touch sensors may be further subdivided into digital or analog or other categories. This listing of sensor types is not intended to be exhaustive, but does give some indication of the many types of sensors available for use on an intelligent robot. It is also interesting to note that most of the sensors are available in "hardened" form for use in environments of high temperature, pressure, and radiation, or corrosive, zero gravity, or other unusual environments.

5.2 Image Processing for Robot Vision

Let's now concentrate on one special sensor, the camera, and investigate what has been and may be possible through the use of computer vision. Computer vision research, development, and applications are aimed at understanding the complex visual processes so that simple, useful solutions to practical problems may be developed. Such areas as cognitive psychology, image processing, pattern recognition, machine intelligence, computer graphics, computer systems architectures and programming languages, engineering, science, mathematics, and even neurophysiology share common interests in the computer vision discipline.

Vision may be used in several ways in an intelligent robot. For accuracy, it may locate and track the robot hand to provide feedback control. The location, orientation, and recognition of parts to be picked up is another important application. Vision may be used to guide a seam-welding robot or to control the mating of two parts. Regardless of the application, the vision system must contain an illumination source, a camera system, and a computer interface. If ambient lighting is relied on for the illumination source, the imaging process may be called passive. This type of imaging is often used in military applications since the position of the viewer is not compromised. In industrial applications there is no such concern, so that controlled illumination or active imaging may be used. The camera system contains not only the camera detector but also, and very importantly, a lens system. The lens determines the field of view, the depth of focus, and other factors that directly affect the quality of the image detected by the camera. Novel techniques, such as using a fish-eye lens to obtain a 360-degree field of view with no focus adjustment, have recently been investigated and appear very useful in mobile robot applications.

The type of camera used is also important. A very popular type that has been used in several intelligent robots is a solid-state camera. This device offers greater sensitivity and ruggedness, is lightweight, and is easily interfaced to a computer. The camera may also contain important processing electronics. The camera/computer interface is also very important, since such factors as the number of resolution elements and distinct gray levels directly affect the quality of the image.

Several companies specialize in robot vision systems and are constantly improving their hardware and software. Two important hardware specifications of a computer imaging system are the number of points per square inch and the number of gray levels stored at each point.

If you consider the general scene, such as the one in front of you right now, you may see a set of objects sitting in some stable position on some supporting structure. Each object may be quite complex in both its three-dimensional surface shape and in its material properties. Just describing the scene so that it can be represented in computer data is a difficult problem in computer vision. The computer specialty that concentrates on displaying scenes is called computer graphics. It is now recognized that a complete representation of even a grain of salt is impossible even with our largest computer unless some form of model or pattern is used.

One way to model or pattern a scene is to break it into structures, then organize these structures into some sort of hierarchy. For instance, the structure of a scene can be thought of quite nicely in terms of a hierarchical tree, which is generally drawn as an inverted tree of several levels. In this case, the scene as a whole is the trunk, the objects that comprise the scene are the main branches, and the properties of those objects are smaller branches and twigs coming off the main branches.

To describe the surface of an object, let's imagine a "blocks world" in which the surface of each object is flat. We call these polygonal objects. Boxes, books, and houses fit quite well into this category of block objects. In the blocks world, each object is composed of a set of surfaces. For example, a cube has six surfaces. To represent this in our computer tree model, we let the three-dimensional cube be a main branch of the tree, the two-dimensional surfaces be small branches coming off the main branch, the one-dimensional properties of these surfaces be twigs coming off the small branches, and the zero-dimensional properties be the leaves.

Since each surface makes up one of the sides of the object, each surface must be represented by a separate, small branch connecting at the main branch. At this level of the tree, we are no longer dealing with three dimensions, but only two—length and width. To continue the decomposition, we note that each surface is bounded by straight lines or edges and is made of some material. These straight edges, being properties of the object's surface, are represented by twigs coming off the little branches representing each surface, and the material information is saved as a label on the trunk. This allows the computer to understand that each surface is bounded by straight lines or edges. At this level of the tree, we have only to deal with one dimension—length. As the final step in this decomposition, we may note that each edge is simply a line between its end points, which we will call vertices. Since the edges are represented by twigs, the vertices are now represented by leaves, and the computer understands that the vertex is an end point of the edge. This final dimensional reduction has taken us from one-dimensional to zero-dimensional entities, the vertex points.

In the real world, the situation is of course more complex because of the curved nature of most natural and many manufactured objects. These are much harder to break down into properties and represent. However, there is a mathematical theory of snowflake curves or "fractals," which introduces the complexities of even block-type objects. For example, a snowflake is a very complex block object. This theory also introduces the idea of a curve that can cover an entire surface and has a dimension greater than one but less than two. These fractional dimensional curves are only one interesting complexity. Another is related to the definition of a curved surface or even just a curve. Admittedly, we know about many types of curves, but how one represents the arbitrarily shaped curved object is still largely a puzzle. For now, we will assume that the surfaces encountered can be adequately represented by combinations of flat, quadric, and spline surfaces. Using all three types, scene description by hierarchical trees can still be made with only slightly more complexity. As we divide up a surface, we must segment it into simple surfaces, which are those that can be represented by planar, quartic, or spline surfaces. Also, for each surface the boundary now may be curved and perhaps described by data points or polynomial coefficients. This description is more realistic and more complex, as we might expect.

Given this method of describing a surface we may now consider two very important problems in the use of intelligent robots. The first is supplying the robot with a model of the scene with which it is working, that is, telling the robot what to expect. The second is enabling the robot to sense its environment and construct the scene so that it may compare sensed information with its modeled information. The first problem is in the province of computer graphics or computer-aided design. The second is the main emphasis of the robot vision field. Both may also be of interest in artificial intelligence, and both may need to be considered for some intelligent robot applications.

5.3 Robot Vision Specification

Let's now concentrate on the robot vision potential. Robot vision systems make the intelligent robot a useful machine in many applications. We are now in the third generation of robot vision systems. The first-generation systems work with silhouette images of objects and infer such parameters as the location, orientation, and size of an object from the shape of its silhouette. These systems are characterized by binary or two-level image processing, with the images produced by a back-lit scene. The second-generation robot vision systems use several gray levels to characterize objects. These systems can work with front-lit scenes and may differentiate texture patterns. The third-generation systems measure not only gray levels but also use stereo techniques to determine the three-dimensional coordinates of the visible objects in a scene. More advanced systems can even infer some information about surfaces that are not visible, such as the back of an object. We will now present some examples of each type of system.

There is an overall common structure between the human visual system and a machine vision system. One powerful computer peripheral that has only recently been made cost-effective and adaptable is the vision system. With computer vision systems, the capabilities of computer perception are greatly enhanced, and its potential for the performance of intelligent tasks is more apparent.

For robot applications, there are two main divisions of vision techniques. The first is for robot control, as illustrated in Figure 5–2. The special requirements of this application are high-speed computations required for robot control; therefore, the techniques used must be elegantly simple. Another class of robotic vision applications relates to inspection for quality control, as shown in Figure 5–3. Although there is still a desire for high-speed computations, the speed may be that of a conveyor rather than that of the robot. The situations are often stereotyped so that more sophisticated techniques may be used.

Illumination Systems

One of the first considerations in robot vision applications is the type of illumination to be used. In many other imaging applications, natural or "ambient" lighting is used. However, for industrial applications, ambient lighting is rarely sufficient. Therefore, an additional illumination system must be selected. Point, line, or area sources may be used. Also, spectral illumination may be selected to provide high contrast between the desired objects and background. Polarizing filters may be required to reduce undesirable specular "glare." Also, since a moving object may be involved, a "strobe" may be required to eliminate blurring resulting from motion. Various types of illuminators have been used with robot vision systems. Light tables that permit back-lighting are excellent for silhouette imaging. Line illumination produced with a cyclindrical lens is a key element in some robot vision systems. Laser point illumination is used in many three-dimensional measurement systems. The selection of an illumination system for robot vision is analogous to the selection of a staining method in microscopy. The proper selection often makes the problem much clearer.

Camera Positioning, Focus, Zoom, and Aperture Control

The selection of a method for camera positioning is also important. A fixed position is the simplest and is appropriate for some object recognition and inspection systems. In other applications, such as seam welding, a robot hand mounting is required. Often aperture, pan, tilt, zoom, and focus servo controls are required. The selection of these systems is application dependent and requires careful study.

Camera Selection

Camera selection is also an important consideration in robot vision systems. One cannot obtain reliable and accurate data from a camera that does not have reliable characteris-

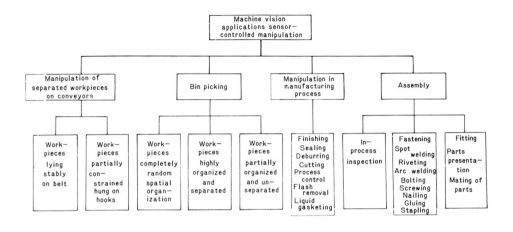

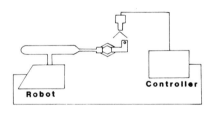

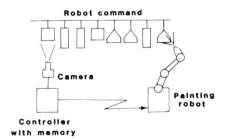

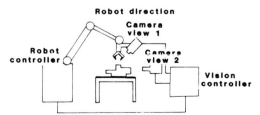

Figure 5–2. Machine vision applications for robot control. (Courtesy of Copperweld Robotics, Inc., Troy, Michigan.)

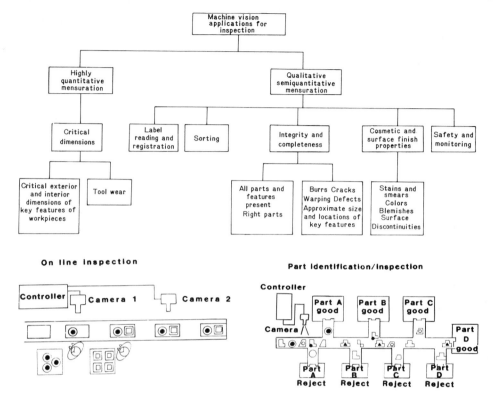

Figure 5–3. Machine vision applications for visual inspection. (Courtesy of Copperweld Robotics, Inc., Troy, Michigan.)

tics. Of the many different types of electronic camera sensors available, two are predominantly used in robotics applications—the vidicon, and the solid-state array. A vidicon is an electronic tube consisting of a glass faceplate with an embedded phosphor surface that is sensitive to light and has an electronic readout capability. The solid-state array camera is an integrated circuit array of discrete photodetectors with associated circuitry. The low cost of broadcast TV cameras often provides a false sense of cost for a measurement-quality camera. Only those cameras produced in large quantities are available at low cost. All these cameras are designed for image visualization, not image measurement. Therefore, a high-quality or special-purpose image-measurement camera can be quite costly. Camera specifications include scanning format, geometric precision and stability, band width, spectral response, signal-to-noise ratio, automatic gain control, and gain and offset stability.

Camera selection includes selection of the lens and camera, digitization and processing unit, and computer interface. Each of these components is important to the overall system. The lens determines the field of view and depth of focus. For a fixed camera location, these are not difficult to determine. However, for a robot hand-mounted system, some method of automatic focus may be required. A variety of cameras is available. A primary consideration is response time. With standard imaging cameras, a frame time of one-sixtieth of a second may be the smallest time increment involved. Tracking cameras are also available that provide continuous x and y locations of a spot. The digitization and control unit may have the capability of digitizing and storing an entire image frame in one-sixtieth or one-thirtieth of a second, or it may require up to several seconds to digitize a frame. The computer interface and processing software is the final element. Although a considerable amount of image processing software has been developed (Hall, 1979), only a small subset of this can be applied to real-time robot control with a general-purpose computer. New computer architectures are being developed for these new applications.

Digitizing an Image

The first step in the processing of an image for robot control is to convert the image into a computer-compatible form. This requires a two-step procedure called sampling and quantization. Sampling is the process of converting the continuous spatial information to discrete sample values at points on a usually equispaced grid. Sampling is usually done at resolutions that are powers of 2, such as 128 by 128, 256 by 256, or 512 by 512 picture elements, or pixels. Next, the brightness or intensity at each point is divided into discrete levels, such as 2 for binary images or perhaps 64, 128, or 256 for gray-scale images. These two steps provide an image that can be stored in computer memory for processing.

Note that the memory size required for storing an image grows quickly as the resolution is increased. For example, the memory required for a 64 by 64 binary image is 4096 bits, which is 4K of memory (1K is 1024 bits). This image could be stored in only 512 bytes since a byte is a group of 8 bits. The same size of spatial image stored at 8 bits/pixel for 256 shades of gray would require 4K bytes of memory. Increasing the spatial resolution to 256 by 256 pixels increases the memory requirements to 64K bytes for an 8-bit/pixel image. Note that this much storage would require an address length of 16 bits and would totally fill the memory on a microprocessor that has a 16-bit address length. Increasing the spatial resolution to 512 by 512 pixels again with 8 bits/pixel requires 256K bytes of memory. This resolution could not be stored in the direct memory of a microprocessor with a 16-bit address length but could be easily accommodated by one with a 32-bit address.

These large memory requirements are one indication of the complexity involved in image processing. A modest microprocessor could easily store several 64 by 64 binary images that require only 512 bytes of storage, but could only hold 16 images of this size at 8 bits/pixel. This same microprocessor could not accommodate a single 256 by 256 image at 8 bits/pixel without using the entire 64K memory.

Advanced architectures specifically designed for high-speed image processing are available; however, these have not been specifically designed for robot vision. The market is still young and growing, so we can expect to see many more robot vision systems develop over the next few years.

Another consideration for robot vision is processing speed. The U.S. standard TV image rate is 30 frames per second. A "frame grabber" that can digitize and store an image in one-thirtieth of a second is required for high-speed processing. Total processing times of the order of one-tenth of a second may be required to provide updates to a robot controller, which is at least 10 times higher than the robot's natural resonant frequency or that required to keep up with a moving conveyor line. More time may be available for inspection applications; however, if a conveyor is moving at 100 inches/second, 10 inches of material will pass by in one-tenth of a second. Processing speed is often crucial.

One image-processing operation that can easily be performed in hardware in one frame time is the computation of a histogram. A histogram is simply a tabulation of the number of image points at each particular gray level. For example, if an 8-bit/pixel, or 256 shades of gray, image is obtained from an analog-to-digital conversion of the camera's video signal, then only 256 storage cells are required to store the histogram. The histogram is computed by simply incrementing the storage cell for each pixel, using the gray level value to point to the correct memory location. A gray-level histogram is shown in Figure 5–4. If an image is to be converted into binary form, the valley in the histogram may be located and used as a threshold value. Any gray level below the threshold is set to 0, and any value above the threshold is set to 1 to produce a binary image.

Robot vision systems greatly enhance the abilities of both computers and robots. Many robotics systems are trained to perform specific tasks and require no human supervision once the training is complete. However, the training process alone can only be used if the task is well defined and does not change. This requires that the environment be well controlled and synchronized with the movements of the robot manipulator. If, for example, a robot is trained to pick up and place a part for assembly, each piece must be positioned in the proper place so the robot can locate it. Robot vision systems can reduce or eliminate this constraint by allowing the robot manipulator to locate each piece by itself. This enhancement to robotics can be used to provide a feedback path to adjust the trained procedure in accordance with the environment.

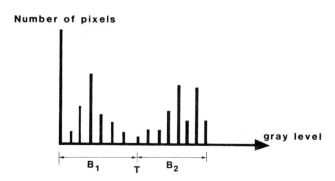

Figure 5–4. Gray-level histogram used to determine a threshold for producing a binary image.

Processing Examples

Some early vision systems located objects by interpreting polygonal surfaces in a block world like the one we described earlier. Objects of this type have distinguishable features that can be identified by segmentation methods. These features can be used as reference points of the surface for object location. However, a featureless curved object, like a ball, has no unique features on its surface, making it difficult to measure from a two-dimensional image. Unfortunately, this class of objects encompasses many machined workpieces.

The three-dimensional solution necessary for curved object location, recognition, and manipulation has many different approaches. The region method of representing polyhedra incorporates an approach called active imaging, which involves a projection system and a camera system. The projection system is used to impose features on the surface of the object, and the many resultant patterns can be used to obtain different types of images. The camera system acquires the object image with the imposed features, and necessary calculations are done to interpret the image for analysis. Active imaging techniques are ideally suited for industrial applications. One advantage of the active approach is that it provides a high degree of control over the environment by providing special illumination or markers for accurate location. The accuracy of the object location derived from a projected grid system can be controlled by the grid spacing. "Stereo" imaging with a single image may also be accomplished. This method uses the projected pattern as the first image and the received image as the second, and a standard stereo solution results. Other techniques involve the use of cooperative algorithms to determine the orientation from a single view using a single camera to recognize overlapping parts, or using geometric and relational reasoning to recognize a three-dimensional object from a single view.

Successful developments in active imaging techniques include the CONSIGHT system developed by General Motors Technical Center (Holland et al., 1979). The CONSIGHT system uses a linear array camera and two projected light lines focused as one line on a belt, as shown in Figure 5–5. The system handles a wide class of manufactured parts in a nonideal industrial environment. The principal functions of the CONSIGHT system are object detection, position determination, part pickup, and part transfer to a stacking location. The overall system may be considered in two parts: the vision subsystem and the robot subsystem. The vision system uses a linear array camera with 256 elements focused on the conveyor belt. It also uses two line beams formed by cylindrical lenses to project a line on the conveyor belt directly under the camera. When an object comes under the camera, a portion of the line is shifted horizontally in proportion to the height of the object. The camera control senses the edges of the object by the absence of the line. Features can then be computed from the edge image, including the position of the object, the area, a bounding rectangle, the area of the largest hole, and the centroid and orientation of the largest object. Recognition is accomplished by comparing the computed features with those of prestored prototypes. Given this information, the robot subsystem is then able to track and pick up the object and place it in an appropriate location. The position on the part at which the robot can pick it up is also determined by the vision system.

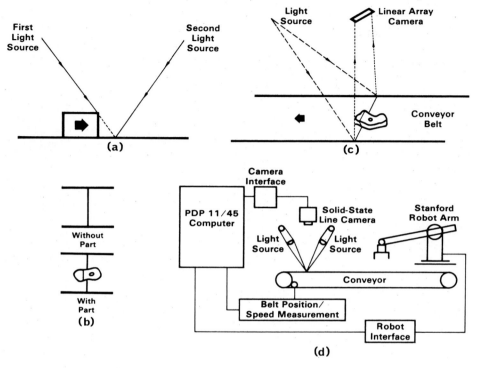

Figure 5–5. The CONSIGHT system developed at General Motors Technical Center. (a) Basic lighting principle. The principle of the lighting apparatus is illustrated here. A narrow and intense line of light is projected across the belt surface. The line camera is positioned to image the target line across the belt. When an object passes into the beam, it intercepts the light before it reaches the belt surface. When viewed from above, the line appears deflected from its target wherever a part is passing on the belt. Therefore, wherever the camera sees brightness, it is viewing the unobstructed belt surface; wherever the camera sees darkness, it is viewing the passing part. (b) Computer's view of parts. (c) Improved lighting arrangement. Unfortunately, a shadowing effect causes the object to block the light before it actually reaches the imaged line. The solution is to use two (or more) light sources, all directed at the same strip across the belt. When the first light source is prematurely interrupted, the second will normally not be. By using multiple light sources and by adjusting the angle of incidence appropriately, the problem is essentially eliminated. (d) CONSIGHT hardware schematic. Since its speed is neither constant nor predictable, the belt is provided with a position and speed detection device. Position and speed information are necessary because the camera scans the belt at a constant rate, independent of belt speed. For each equal increment of belt travel, the vision subsystem records one of these scans. Belt travel increments must therefore be measured precisely. (e) A realization of the CONSIGHT system on a robot assembly line. (Courtesy of General Motors Research Laboratories, Warren, Michigan.)

120 | Robot Sensors

Figure 5–5 (continued)

(e)

Another system was developed by Albus at the National Bureau of Standards (Albus, 1981). Albus's system also uses a plane of light to determine the position and orientation of parts on a table. However, the camera and a strobed line light source are mounted on the robot arm. This gives the system the flexibility of approaching a set of objects from many different directions. Also, the camera used is a two-dimensional array so that the amount of displacement in the line may be used to determine the distance to the object. The line projection technique works quite well on locating the edges of an object but requires scanning to locate all the object surface points.

The use of grid coding for image segmentation based on the spatial frequencies of the projected images has also been proposed. Other computation techniques have been used recently on the laser/shutter/space encoding systems. These systems use the time and space coding of dots to measure surface coordinates.

The human visual system features two perspective views of objects from which distance or depth information is derived. A computer vision system can do the same thing by using two cameras to view an object from two different perspectives. The human visual system also has the ability to detect depth changes from shading information. Depth information can also be derived from images obtained by a computer vision system through an analysis of the shading of the surface along with a knowledge of lighting conditions. A model can be obtained, given a format for the model, using numerical or analytic techniques of least-squares curve fitting. Once a model is obtained, the mathematical model may be decomposed into a qualitative description of the surface. The model could then be used to describe the power of the return signal as a function of the incidence angle between the surface normal and the transmitted signal vector. There is a concern for characterizing the reflection of light on surfaces with different orientations. Numerous computer-aided design systems, such as those marketed by General Electric, Computer Vision Corporation, and Applicon Systems, can simulate the appearance of a mechanical part or architectural structure through graphic reconstruction. For these reconstructions to appear natural, an understanding of the reflection characteristics of surfaces is required.

Polygonal surfaces are among the simplest to describe. Simple techniques are used in their reconstruction on graphic displays. One popular program, called MOVIE.BYU, developed at Brigham Young University (Christenson, 1978), uses a vertex list and a vertex connection table to describe a surface. This representation provides sufficient information to reconstruct the projection of the surface onto a plane positioned anywhere in space and oriented in any way relative to a global coordinate system. Other data structures for describing polygonal surfaces use node or vertex locations with a variety of schemes for describing the connection of points in graphic reproduction. However, similar techniques cannot be applied to curved surface representation since there exist no vertices, edges, or sides on a curved surface, such as a sphere.

Polygonal surfaces are constructed of multiple intersecting planes, each of which may be described by a linear equation. The line defined by the intersection of two or more planes is considered an edge, and a vertex is the point of intersection of two or more edges. The edges of a polygonal object are discontinuities in the surface description. Since an object of this type is constructed of bounded planes that form the polygons, the edges are line segments at which one element of the surface ends and another begins. Curved surfaces may also have edges formed by discontinuities if the surface is formed by multiple bounded curved surfaces where each of the elemental curved surfaces describes only a bounded region of the entire surface. With no restriction on the order of the equations required to describe a general surface, the complexity of the mathematical model could approach that of the molecular-level description. Again, the search for an accurate description of a general surface becomes futile. Therefore, the methods employed in the representation of arbitrary curved surfaces must rely on approximation techniques. Many manufactured objects and structures can often be described by low-order equations for which a mathematical model can be derived. In such cases, a mathematical model would be the preferred method of representing the surface, since a single equation may be used to obtain any point on the surface. Numerous methods of describing curved surfaces, both by mathematical models and by approximations, are known.

A surface recognition technique requires a sufficient sample of surface points, as required by the sampling theorem, for adequate surface description. Stereo vision has long been accepted as a valid method of surface measurement; however, since the stereo vision technique requires the identification of corresponding points in the two images, this method is difficult to apply to curved surfaces. The vertices of polygonal solids are the features most commonly used to match two points on a pair of images. Curved surfaces may not contain vertices, but still be closed surfaces. If a featureless curved surface, such as a sphere, is viewed from two different positions, the two images may appear identical under uniform lighting. One solution to curved surface measurement utilizes the characteristics of the change in surface reflectivity as the illumination source is moved from one position to another.

The models that are applied to image synthesis suggest the ability of the vision system to recognize surface shapes from shading information. The process of surface shading involves the assignment of an intensity or color to every picture element in the image that accurately simulates the viewing situation. The shading of a surface point depends on the surface reflection characteristics, the surface geometry, and the lighting conditions. Each of these properties must be considered in the development of a surface shading model. Models used in image synthesis for the shading of the surface of an object can be applied to image analysis to obtain the surface shape from the shading information. However, the ability to perform this task with sufficient accuracy depends upon the ability to select an appropriate shading model that closely approximates the reflection characteristics of the surface material. The problem of selecting an appropriate shading model is a topic that requires continued research. Present techniques of calibration are difficult to apply to automated systems, since a new set of calibration parameters must be obtained for every surface to be examined.

The technique of stereo vision for surface measurement is well known in photogrammetry and has long been applied to three-dimensional data acquisition and object description. The stereo vision process requires that the camera models be known for each view of the object. Through a knowledge of the camera model parameters of location, orientation, and focal length, the perspective projection transformation matrix for each camera can be determined. Since it is often difficult to obtain the model parameters directly, the camera calibration procedure may be applied using six points for which the global coordinates and image coordinates are known. Once the transformation matrices are known, the global coordinates of a surface point may be computed.

Curved surface representation techniques have been described whereby a model of the surface may be obtained that is applicable to surface recognition. These models may consist of mathematical models or numerical approximations. To obtain the necessary sample surface points to describe the surface geometry adequately, several methods of curved surface measurements have been developed. One method is based upon the relations between surface shading and surface geometry. A second method utilizes the stereo vision approach to surface measurement but requires only a single image of the surface to be analyzed.

Curved surface representation is a necessary step in curved surface recognition. Because objects comprised of curved surfaces may have no edges or vertices, surface models must be used to describe their geometry in three dimensions. These models may consist of mathematical models or numerical approximations, such as splines. Mathematical modeling is often the best approach to surface description, since the information contained in the model may be easily decomposed into qualitative descriptors, including shape, size, orientation, and location. A description of this type provides an object recognition technique independent of the location or orientation of the object.

To describe an object, some level of knowledge must exist for which a qualitative description can be derived. The level of knowledge obtained by the two techniques to be considered is a set of sample surface point coordinates. One method of obtaining sample surface points is by examining the shading of the surface under controlled lighting conditions. It has been demonstrated that the models used in image synthesis for surface shading can be applied in image analysis to obtain the surface shape. If an appropriate shading model for the surface material and texture is obtained, a measurement of the surface normal vector for the surface's points viewed by each pixel in the image can be computed. This method does not require a high-resolution image to accurately obtain a sufficient number of sample surface points to describe the object. Obtaining an adequate shading model for a particular surface material and texture is a difficult problem that will require further research.

A second method is described that utilizes the stereo vision principle. Vertices and edges are the features most commonly matched in a pair of images, and by triangulation, the three-dimensional location of the feature is obtained. Featureless curved surfaces may have no edges or vertices and still be closed surfaces. The solution to this problem is to impose features onto the surface through "active imaging." The system described uses both a camera and a projection system. Because the projection process is identical to the imaging process with respect to the process models, this system performs stereo vision with a single image (Hall et al., 1982).

5.4 Commercial Robot Vision Systems

Three different types of robot vision systems for robot control and inspection are commercially available. The first type deals with two-dimensional binary images, or silhouette imagery. Many of these systems are based upon the pioneering research done at SRI International in Menlo Park, California. The second type uses gray-scale imagery from which more discrimination is possible but for which longer computation times may be required. The third type uses some type of structured light and stereo triangulation to determine the three-dimensional surface of objects. Each approach requires a lighting system for illuminating the objects, a camera for gathering the image, and a storage device for recording the image during processing.

Silhouette Robot Vision

As an example of a silhouette robot vision system, let us consider the Machine Intelligence Corporation's VS-100 system. The system may be divided into two major parts: a camera processor and a feature processor, as shown in Figure 5–6. The camera processor can control a strobe illuminator and accept input from one or more cameras. The LSI-11 microcomputer performs the supervisory tasks, computes features, and performs the object recognition. The system recognizes objects by comparing features on the test object with those of stored prototypes. The prototype features are determined during a training session in which exemplars are shown to the system and the recognition features determined. Several features, including geometric properties, such as the area

124 | Robot Sensors

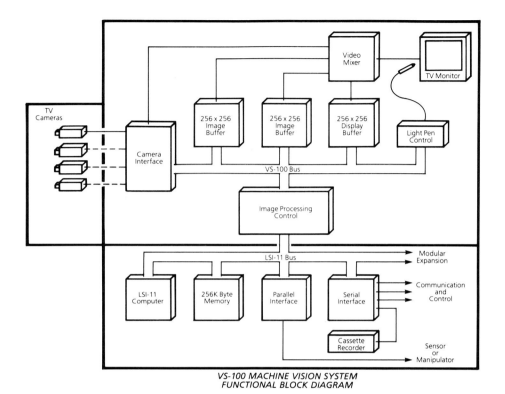

Figure 5–6. Machine Intelligence Corporation's VS-100 system functional block diagram showing divisions into camera and feature processor. (Courtesy of Machine Intelligence Corporation, Sunnyvale, California.)

and perimeter, as well as orientation properties, such as the centroid and principal orientation, may be selected for computation. Note that the geometric features are invariant to location and orientation and permit recognition invariant to the factors.

During training an image is scanned, a histogram of the gray levels is computed, and a threshold value is selected. The threshold is usually at the valley between the light and dark peaks in the histogram. If a clear valley does not occur, this indicates that the lighting should be adjusted to produce a sharper contrast between the object and the background. One illumination method that produces a very sharp contrast is backlighting the object. This is easily accomplished using a light table. High contrast can also be achieved in most cases using front lighting with an intense source.

Recognition of objects involves several steps. First, the image is scanned and, using the predetermined threshold, is stored in binary form. Next, the outlines of the objects are determined by finding edge points. Starting at the upper left-hand corner and moving line by line from top to bottom, the edge transition points are located. A compressed image consisting of the length of pixels of the same brightness or run length code is also determined. Whenever an edge point is determined, the computer determines which object it belongs to by examining neighboring points. It then enters this edge point into the list of edge points for that object. The edge lists are considered tentative until the scan is completed so that objects with holes or protrusions may be accommodated. The edge lists are processed at the end of the scan and merged if necessary.

After the edge lists have been processed, the features, such as the perimeter length and area, are computed, as well as the location and orientation. Finally, the system compares the computed features with the prestored prototype features to recognize the object. An illustration of the processing and the VS-100 system is shown in Figure 5–7.

As another example, let's consider the first vision system offered by a robot manufacturer. The Copperweld Robotics, Opto-Sense vision system could be used alone or interfaced to a robot. Typical applications have included measuring automobile frames on a moving assembly line, looking for missing parts in metal subassemblies, detecting missing objects in packages prior to sealing, and inspecting complex stamp-

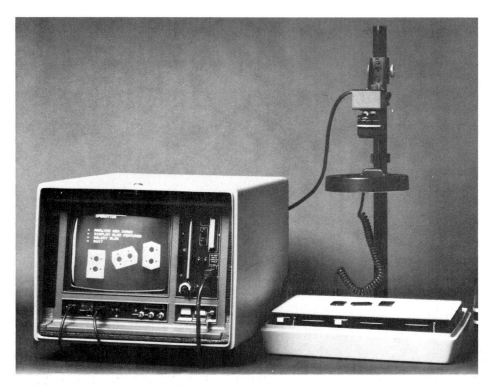

Figure 5–7. The Machine Intelligence Corporation system showing operation and images. (Courtesy of Machine Intelligence Corporation, Sunnyvale, California.)

ings for the presence and location of holes and other details. When used with a robot, the system provided the robot with visual feedback. The Opto-Sense system consisted of a computer controller that controls one or more solid-state cameras, such as the General Electric TN 2500. The image was divided into an array of 244 by 244 pixels. Software was usually tailored to particular applications using a subroutine package for various tasks.

In one application of robotic inspection at the Chevrolet Motor Division of General Motors in Flint, Michigan, the Opto-Sense was used with four cameras to determine if valve covers for engines were properly assembled. The cameras were mounted overhead in protective enclosures with transparent windows. Since the cameras were focused on the assembly line, dust on the lenses had very little effect. Reflected light and shadows permitted the imaging system to discern the presence or absence of detail. The first task performed by the vision system was to establish part identity, that is, whether a left- or right-side cover was being inspected. The second task was to determine the presence or absence of all necessary characteristics, such as clinch nuts, metal brackets, baffles, and holes. This step also determined whether any extra parts were added. This information was analyzed, and an accept or reject decision was made. This decision information was then given to the robot controller, which also had information available from other tests. The robot controller then determined whether the part should be put into an accept chute, a visual reject chute, or a leak test reject chute. The robot then put the part in the appropriate location. The system increased production from 300 to 1200 parts per hour. This was a 400 percent increase in productivity with 100 percent inspection. The system diagram is shown in Figure 5–8.

Gray-Scale Robot Vision Systems

As an example of a robot vision system capable of gray-scale processing, let's consider the Control Automation, Inc., InterVision 1000 system, shown in Figure 5–9. This system processes 6 bits/pixel, or 64 gray-level images, at rates up to 10 per second on the entire image or 80 window inspections per second.

Another robot vision system capable of processing gray-scale images is the Automatix, Inc., Cybervision III system, shown in Figure 5–10. The Automatix system is based upon a special hardware architecture using 68000 microprocessors and is programmable in a high-level language called RAIL.

Three-Dimensional Robot Vision

A striking example of three-dimensional robot vision is the system made by Robotic Vision Systems, Inc., shown in Figure 5–11. This system was designed for guiding a seam-welding robot and consists of a light projector and camera mounted in protective covers on the robot wrist. The system measures the orientation, position, width, and depth of a seam to be welded. This information is then transmitted to the robot controller to guide the robot path and weld parameters. This eliminates the need for expensive special-purpose fixtures to precisely locate the seam with respect to the robot. Other

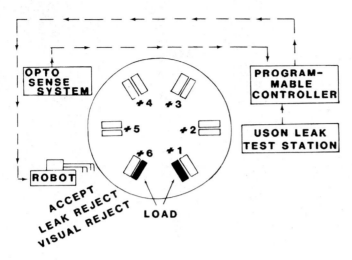

Figure 5–8. Copperweld's Opto-Sense inspection sequence diagram. (Courtesy of Copperweld, Inc., Troy, Michigan.)

INSPECTION SEQUENCE:

St. No. 1: (left-hand fixture) load
St. No. 2: Uson Leak Tester
St. No. 3: Idle position
St. No. 4: Opto Sense System
St. No. 5: Auto-Place Robot sorts and unloads valve covers
St. No. 6: (right-hand fixture) load

applications of the system include automatic control of robot grinding, assembly, and materials handling and routing; automatic inspection of cast, forged, and machined parts; inspecting for missing parts; and three-dimensional copying and scaling of parts.

5.5 Range and Proximity Sensors

A range sensor measures the distance from the sensor to an object. Both point or proximity sensors and range image sensors have been developed for robotics applications. One of the most popular ranging devices is the Polaroid ultrasonic device originally developed for focus control of cameras. This device is now used on many robots, especially mobile robots for obstacle avoidance. A control computer generates an initiate signal to the Polaroid ultrasonic electronics control board. The ultrasonic board then generates a series of 56 pulses of four different ultrasonic frequencies for a total duration of 1 millisecond. This signal is applied to the transducer and can be heard

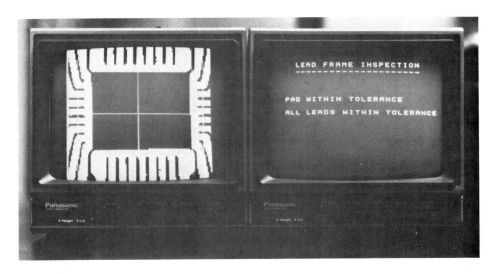
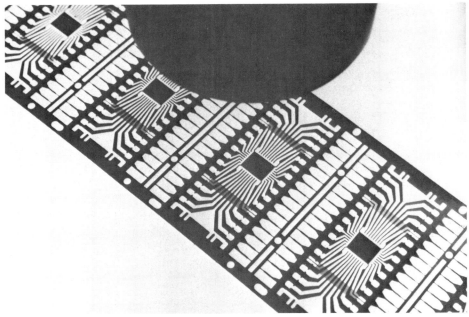

Figure 5–9. Control Automation's InterVision 1000 vision system is a computer-based inspection system with solid-state camera that combines binary and 64-level gray-scale processing to inspect products at 4800 parts per minute. InterVision can be used for noncontact measurement, profile recognition, and presence or absence inspections of randomly oriented, moving, or fixtured products, such as lead frames, leads of components, and populated printed circuit boards. (Courtesy of Control Automation, Inc., Princeton, New Jersey.)

Figure 5-10. Automatix, Inc., vision system, the Cybervision III system. The system is programmed in a structured language that incorporates special vision features and is easily interfaced with some robot controllers. (Courtesy Automatix, Inc., Billerica, Massachusetts.)

as an audible "chirp." A pulse-sent signal is then sent to the computer and control circuitry at the same time that the pulse is transmitted. This signal causes the range count to increment at a rate of 3.2 kilohertz. Upon detection of the reflected echo, the Polaroid board sends an echo-received signal to terminate the count. A maximum range count of 127 corresponds to a distance of 22.5 feet. A single count represents a distance increment of about 2 inches. The range represents the minimum range encountered in a cone of sensitivity corresponding to the radiation pattern of the ultrasonic beam.

Laser range finders and imaging devices have also been developed for robotics applications. For example, the time of flight can be measured directly for large distances. However, since light travels at a speed of 1 foot per nanosecond, very high frequency counters are required. An alternative approach, which is more appropriate for the distances encountered in industrial applications, is to measure the phase shift between the transmitted and received beams. Another approach is to use the laser as a light source in a stereo triangulation system to measure the distance to an object. One possible problem with laser devices is the eye hazard to humans. They must be used with care.

5.6 Tactile Sensors

When a robot hand is close enough to touch an object, contact sensors may be used. Just as in the human skin, which contains pressure, pain, and temperature sensors, a variety

130 | Robot Sensors

Figure 5–11. Robotic Vision, Inc., three-dimensional vision system mounted on a robot and used for guiding a seam-welding gun. (Courtesy of Robotic Vision, Inc., Melville, New York.)

of contact sensors for robots have been developed. Force or pressure, which is simply force per unit area, can be measured at a single point to help adjust the position of a tool, or at several points to provide enough information for the robot to recognize an object.

Perhaps the simplest sensor in this category is a touch sensor implemented with a microswitch. Such sensors are often used to stop the motion of the robot in a particular axis or direction. For example, the feedback on a nonservo robot may be supplied by such a stop or switch in a "move until touch" mode. Also, the closing of a gripper may be halted when a switch is closed. Other applications include sensing that a target has been reached, such as in spot welding, preventing collision, centering the robot gripper on a object without moving it, measuring object dimensions using the switch in conjunction with the high-precision joint encoders, and determining object presence. Point touch sensors are excellent for such tasks in which a single point of contact is sufficient. They are often inadequate for such tasks as part placement or recognition in which several contact measurements are needed.

Force sensors employ a transducer, such as a piezoelectric element, which provides a signal proportional to the deflection and therefore force applied to the point of contact. Such a measurement may be used to provide force feedback for collision recovery or to permit a robot gripper to grasp a delicate object, such as an egg, without crushing it.

In many assembly applications, a sensor that can measure the three-dimensional forces and torques may be very useful. For example, in inserting a bearing into its holder, a very high tolerance may be required. A single force measurement would only indicate that the bearing is in contact with the receptacle. However, three force measurements in the x, y, and z directions could indicate the displacement of the part from the hole. Additional measurements of the three torques (force times distance) may be sufficient to also yield rotational offsets that may be used with the displacement information to permit insertion of the part. Astez Corporation makes such a force-torque sensor that may be attached directly to a robot wrist.

Arrays of touch or force sensors may provide sufficient information to recognize an object, determine how the object is resting, determine its center of mass or pickup point, or its orientation. Lord Corporation of Erie, Pennsylvania, has used such an array sensor for object recognition and location for assembly. The Lord LTS-100 tactile sensor is 3.18 inches square and 1.12 inches thick. It contains 64 sensing sites mounted on an 8 by 8 array with 0.3-inch spacing. The device employs two stages of transduction to convert the impressions on its touch surface into easily processed electrical signals. One stage is a mechanical deflectometer that has an elastic compliant touch surface in which an 8 by 8 array of pins constructed of the same elastic material are embedded. A deflection of any pin is measured by a light emitter and detector pair. As the pin is deflected, it shadows a portion of the light. The maximum amount of light is received by the detector when no deflection is present, and the minimum amount when full deflection of the pin shadows most of the light. The deflections range from 0 to 0.8 inch at 1 pound force. The sensor light signals are converted into 64 levels of displacement. Since the device measures displacement, force measurements are developed by comparing the amount of deflection to the force deflection characteristics of the elastic material. The sensor has been used in conjunction with a Unimate 560 robot for assembly of a flexible diaphragm and plastic cap. In this application, the tactile sensor was located on a bench. The robot using a vacuum gripper picks up the part and places it on the tactile array to determine which side is facing the array. If the wrong side is presented, the robot picks up and turns the part. The part is again placed on the tactile array to determine its center location. With this position known, the robot can then place the diaphragm correctly into the plastic holder. Lord's newer sensor, the LTS 200, incorporates several major changes. Its overall size is reduced to 1.75 inches long by 1.125 inches wide and 0.65 inch thick. The sensing array is 12 by 8 on 0.1-inch centers. Each site is sensitive to loads as low as 0.05 pound and up to 25 pounds perpendicular to the touch surface. It also includes a new module that measures gross normal and sheer forces. The Lord system is shown in Figure 5–12.

Another example of a commercial touch-sensing device is the tactile sensor developed by Barry Wright Corporation of Watertown, Massachusetts. The device is called the TS402 and consists of a touch sense pad on an electronic interface that connects to the robot controller. The active area is 1.56 by 1.56 inch square and contains a 16 by 16 array of sensors located on 0.1-inch centers. The overall size is 2.50 inches wide by 2.60 inches long and 0.35 inch thick. The device uses elastomers that provide data that may be used to determine force, position, and part orientation.

A survey conducted by Case Western Reserve University (Harmon, 1982) of tactile sensing tasks needed in industry indicated that many of the desired operations could be satisfied with rather gross sensing using fewer than 10 sensor elements per square inch and a 4 by 4 array of elements. However, at least 7 percent of the applications did require fine tactile sensing, with more than 10 elements per square inch and larger than 16 by 16 arrays. Slip sensing was also determined to be required in about 53 percent of the tasks studied.

One major result of Harmon's study was that the required tactile spatial resolution required for a wide range of assembly tasks was surprisingly small. Arrays of no more than 8 by 8 pressure-sensitive points with interpoint spacing of 0.1 inch and capable of slip sensing appear to promise considerable capacity. Such arrays, mounted on two-fingered grippers and three-fingered anthropomorphic hands, were concluded to be able to replace humans in 80 percent of the assembly tasks studied.

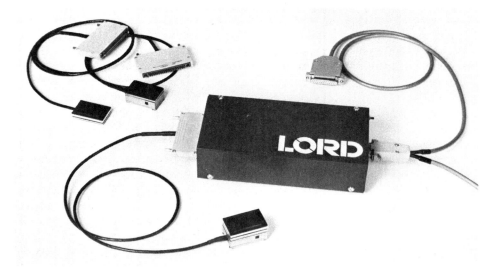

Figure 5–12. Lord Corporation's tactile sensors and interface unit. (Courtesy of Lord Corporation.)

5.7 Sensors for Mobile Robots

Autonomous, self-guiding, mobile robots require special sensors that are not necessary for stationary robots. The purpose of this section is to review the sensor requirements for mobile robots and some solutions for the various problems encountered in using sensor devices. Safety considerations make it most desirable for mobile robots to be equipped with sensory devices that can enable the robot to, for instance, avoid collisions or use sensory feedback information for guidance and position determination and often for target location. These requirements include contact tactile sensors, proximity sensors, local and global position sensors, and level sensors. Many applications for intelligent mobile robots exist. Some of these include industrial carts for material transport, military sentry duties, medical patient care, and domestic duties, as well as lawn mowing and vacuum cleaning.

The "Shakey" robot, developed by SRI International about 10 years ago, is an excellent model for the type of robot we wish to consider (Raphael, 1976). It consisted of a mobile wheeled base, an on-board computer, and several types of sensors, including simple contact switches, a ranging sensor, and a TV camera. Shakey could maneuver through a room, avoiding obstacles and performing simple tasks.

Hans Moravec (1982) of Carnegie-Mellon University (CMU) has developed a camera-equipped mobile robot to support research in control, perception, planning, and related issues. The CMU rover has a cylindrical shape and is about a meter tall and 55 centimeters in diameter. It has a steerable wheel assembly consisting of three independently steerable wheel assemblies. It also carries a TV camera on a pan-tilt-slide mount, as well as several short-range infrared and long-range proximity detectors.

Bart Everett (1982) developed a mobile sentry robot, which he called ROBART, at the Naval Postgraduate School. This robot was designed as a development robot for autonomous sentry applications with emphasis on testing appropriate sensors and their associated interface hardware. ROBART is designed to function autonomously and is equipped with a complex scheme for collision avoidance. Also, the battery condition is constantly monitored by its on-board computer, which can activate a radio-controlled homing beacon on a nearby recharging station when needed. When a low battery condition is detected, the beacon activates and the robot homes in on this station, connects to the charger, and replenishes its energy supply.

Dr. Hall and his students at the University of Tennessee, built two mobile robots. The first, called Micromutt, contained two independent drive wheels, which permitted a zero turning radius, an on-board microprocessor, audio sensors to permit homing, and sonar sensors for collision avoidance. The second robot, called MERV, contained many of the same capabilities but with additional sonar sensors to not only detect an obstacle but also to determine the direction of motion. These robots were built as prototype educational robots and proved an excellent experience for the student designers.

Another class of mobile robots included the popular show and home entertainment types, such as Brains on Board (BOB), which is a 3-foot-tall mobile robot that uses, along with other sensors, an infrared sensor attuned to the wavelength of the human body that permits it to locate people in a room. It is mobile and is equipped with two microprocessors and a voice synthesizer. It is built by Androbot, Inc., of Sunnyvale,

California. Other robots that are currently being offered include HERO I by Heath Company, ComRo by Comro, Inc., and RB5X by RB Robot Company. Common features include on-board microcomputers, ranging sensors, light sensors, sound sensors, and voice synthesizers. A commercially available mobile robot vacuum cleaner is also available as an attachment to the RB5X robot and is offered by The Sharper Image.

Several special sensors are required for mobile robots. Furthermore, since the sensors may be used for controlling a robot that may be moving at velocities in the range of 1 to 20 feet/second, high-speed algorithms may be required. Contact switches, proximity detectors, homing signals, light and sound detectors, level indicators, and local and global positioning devices may also be required.

Contact switches placed around the periphery of the robot provide a last line of protection for collision avoidance. The switches could be used to simply stop the robot; however, if they are arranged properly, a new direction of motion can also be inferred that allows the robot to continue on its way. Contact switches that simply stop the robot when it touches an obstacle are used on the popular mail cart robots. Several switches around the periphery are needed to provide the motion control.

Proximity sensors, such as the Polaroid sonar ranging device, may be used to change the robot's path before it encounters an obstacle. A single sensor is of limited use; however, one proximity sensor may be used to slow down the robot before it encounters the obstacle. Three proximity sensors may be used to provide steering information. For example, if an object is detected in front of the robot, side-mounted proximity sensors may be used to guide the robot in the direction in which it has the maximum unobstructed traveling distance. Peripherally located proximity switches may also be used to provide a safety or warning system for the robot, which tells the robot if something or someone is approaching it. Infrared proximity sensors for intrusion detection have been used by Everett for use on security applications for mobile robots.

One of the most important and difficult sensor systems required for a mobile robot is a position location device. Both local and global position information may be required. The accuracy of this information is also very important in determining the control strategy of the robot, since the success and accuracy of the manipulator's task is directly related to the success and accuracy of positioning the robot. Local positioning information may be achieved by implementing shaft encoders on the wheels. These encoders can provide accurate information for short distances; however, wheel slippage and other factors can cause large errors to accumulate over large distances. Therefore, some global method for determining updates to the position may be required.

Several methods can be used for providing position updates. The LORAN system, used in navigation, uses radiofrequency beacons and receivers that determine the time delay from the known location beacons to permit triangularization of position. Unfortunately, the accuracy of these devices is of the order of tens of feet. A local LORAN-type system could also be built; however, it would require active transmitters as well as receivers.

A global positioning system using modulated mercury arc lamps modulated at different frequencies and a scanning optical detector to measure the angles to the known positions was built into a mobile robot cart (Anbe et al., 1972). Other methods for obtaining global positioning may use passive reflectors or targets. The advantage of a passive approach is that the positioning location device can be located on the mobile robot and powered from the same source.

The use of a global positioning device may also require that a map be programmed into the robot's memory, so that a strategy based upon its current position and desired position may be developed. This realization has led some researchers to develop methods for mapping the robot's environment. For example, a ranging device on the mobile robot would permit the collection of range data to objects in the surrounding area and, with further processing, be used to produce a map.

Example: The PEGASUS System. As an example of the type of sensors and controls required for a mobile robot, the PEGASUS system design developed by Dr. Hall and his students will now be described. The goal of this project was to develop a system for the autonomous operation of a commercial Hustler lawn mower. The system was to be interfaced to the lawn mower in such a manner as to provide safe, autonomous control requiring minimum human supervision. PEGASUS contained both training and automatic modes. A remote control override capability would also be used to ensure safe operation by keeping a human to supervise the operation.

During the training mode, the operator was to drive the robot around the perimeter of the field and indicate any critical obstacles. During the automatic mode, the robot would select and implement a mowing strategy. The sensors would permit responses to changes in the environment, such as avoiding obstacles, updating its position, or responding to variations in terrain and avoiding any unforeseen obstacles. Both local and global positioning systems were used. For local control, shaft encoders provided the basic position data. A local imaging device was also used to permit the mower to follow the grass cut line and adapt to small directional changes. A global positioning system would provide periodic updates to correct the positional data.

Several mowing strategies could be followed. The strip strategy consists of translating the mower back and forth across the field oriented along the greatest length of the field. At the end of the field, a 180-degree rotation is executed to permit the next strip to be mowed. The perimeter strategy consists of outlining the perimeter and mowing in ever-decreasing areas until the field is mowed. A sector strategy consists of dividing the field into sectors and using either the strip or perimeter method for each sector. It is also possible to combine the strategies. For example, the operator could outline the perimeter of the field and major obstacles in the training mode, then let the computer sectionize the field into sectors, prioritize these, then mow each using a strip pattern. With any of the strategies, obstacles can be avoided in two ways. In the training mode, the operator could indicate the location of obstacles by moving the mower along the outer boundaries of the obstacles. This would permit mowing around lakes or large flower beds. During the automatic mode, small obstacles, such as trees, could be automatically avoided using the sonar obstacle avoidance. This would free the operator from locating a large number of small obstacles during training. In the automatic mode, the contact switches would also prevent the mower from getting stuck between obstacles, such as between two large rocks or trees.

An integral part of the fail-safe nature of the design was the perimeter contact switches. They provided a last line of defense in obstacle avoidance. The contact switches completely encircled the mower to permit not only stopping the mower but also determining the new direction of motion when an obstacle was encountered. The switches were arranged at a height of 6 inches above the ground and 6 inches from the mower. The switch information provided interrupts to the drive system and caused the mower to stop, back away from, and then proceed around the obstacle.

Shaft encoders fixed to the hydrostatic drive motors monitored the mower's position, translation, and speed. The x and y location of the mower relative to a starting or home position was determined from the angular information provided by the shaft encoders. Some compensation for wheel slippage, such as that detected by an increase in instantaneous velocity increases, could also have been implemented. The hydrostatic drive motors were controlled by linear actuators attached to a hydraulic pump. The two drive wheels were mounted on an axis of symmetry and encoded, providing phase pulses and directional information. Wheel slippage was detected by comparing the instantaneous angular velocity to a running average velocity. Any increase in the instantaneous velocity over the average velocity greater than a threshold was considered the start of wheel slippage. When the velocity difference became less than the threshold, the end of the slippage was assumed. The time interval of slippage and the average velocity permitted an estimate of the true distance traveled.

The major system for obstacle avoidance was a perimeter array of proximity detectors. The sonar system was an assembly of sonar detectors, such as those used on the MERV project. However, this system was designed to detect the location of objects near the mower in all directions. This permitted the implementation of a 20-foot safety zone around the mower, monitored by the sensors that automatically stopped the mower, and disengaged the blade if someone approached it from the rear or sides, as well as directional control for obstacles approached from the front. The design consisted of 12 sonar transducers and four stepper motors. The stepper motors rotated their respective transducers through 270 degrees on the sides of the mower, and slightly less than 180 degrees on the front and rear. Four of the stepper motors were located at the four corners of the mower. Each of these had two sonar transducers, one to detect low obstacles and the other to detect high obstacles. The other four motors were located on the front, sides, and back of the mower and had one transducer each.

For safety reasons, a radio control (RC) unit was used to permit the operator to shut down the system at any time. This system also permitted the operator to remotely drive the mower. A standard RC controller was modified to interface with the digital computer controller and was used in compliance with the Federal Communications Commission (FCC) rules concerning radio control equipment.

Since small terrain variations may often be encountered, a local control device consisting of a line scan camera was used to provide detection of the grass cut/uncut boundary and corresponding guidance signals to the mower. A 256-element camera was used. The camera was positioned to view the front pathway. The signals from the camera were converted to binary, averaged, and used to determine a sectored cut line vector. The basic detection depended on the lighting conditions and height and texture of the grass. Therefore, a processing step to determine the validity of the derived vector was implemented by analyzing the composite histogram of the image. The magnitude of the vector represented the remaining distance along an orientation of the grass cut. The direction of the vector represented the mower's direction with respect to the field.

A global positioning system was used to provide periodic updates to the system position. The absolute global positioning system used an omnidirectional imaging system implemented with a horizontally mounted fish-eye lens. The system was gimbal mounted to provide a constant horizontal orientation, which simplified the computation of position. The position was determined by imaging two or more fixed location markers. The algorithm used is similar to that previously developed by Fukui (1981). His demonstration used a conventional field-of-view lens to view a standard target with a fixed-height camera. The position of the camera was determined from the view angles of the target in the horizontal and vertical directions. This system's arrangement required the horizontal view angles of two targets to determine the position of the camera and, consequently, of the mower. Targets were selected to provide high contrast with the background and distinguishability from each other.

Imaging from a mobile robot involves several special requirements. First, since it is desirable to image while moving, motion blurring must be avoided. Since it is impractical to use a strobe illuminator during daylight, a shutter mechanism with speeds of 250 to 1000 frames per second is required. Also, since a very short illumination time is required, a camera with high sensitivity is necessary. Finally, high resolution is desirable since the overall accuracy is determined to a large extent by the digital resolution.

The gimbal-mounting system in this example also provides sensors that measured the steepness of the grade. If a grade steeper than the drive grade of the mower was encountered, the mower could stop, back away, and attempt another route. The pitch and roll sensors, mounted on the gimbal rotation axes, provided the level information. This feature prevented the mower from overturning. These sensors also provided information as to the validity of the global position information, since global updates were valid only when the camera was horizontal.

The computer control for the PEGASUS system was a hierarchical architecture. A 16-bit 8088/8087 MPX-16 single-board computer was used as the supervisor. This computer was programmed in the FORTH language. The 8087 math microprocessor supported 80-bit floating-point operations and increased the numerical throughput by a factor of 500. Seven Z-8 microprocessors operated as dedicated controllers for the individual subsystems.

The main computer processing unit (CPU) was used to control the overall strategy as well as serve as the global positioning system interface. The sonar sensors, linear array, and remote control system were all controlled by separate Z-8 microprocessors. Each of the two drive motors and shaft encoders was also controlled by a Z-8 microprocessor. One other Z-8 was used as a steering coordinator and was connected to both the motor controllers. A final Z-8 was used as a serial traffic controller. A total of eight microprocessors were used in the controller.

The computer language selected for the master control program was FORTH. This is a high-level language; however, due to the nature of its compilation and execution, it is faster than most high-level languages and can operate at speeds approaching that of assembly language. A FORTH program is made up of strings of commands, most of which are one-word commands. The feature that makes this language unique and so attractive is that the user can create new commands as needed. The building block nature of this language permits the programmer to define new commands by stringing together a series of previously defined words. After each new word is defined, it is compiled and added to the dictionary of previously defined words. The new word is then available for use, just like the words provided with the system. Words are built of lower level, error-free words, and debugging consists of determining whether the lower level words operate together as expected. In essence, a FORTH program is simply a word that has been defined as a series of lower level words.

5.8 Sensor and Control Integration

Since a variety of sensors may be required for an intelligent robot, the problem of integrating the sensory information with the stored information to develop a control strategy is also important. In some cases a single computer may be powerful enough to control the robot. In more complex systems, a hierarchical, distributed computer can be used by the mobile robot or flexible manufacturing system. An executive controller may be used to implement the overall strategy. It communicates to a series of dedicated processors that control the robot functions and receive input from the sensor systems. Sublevels in the hierarchy may also be used for related tasks. A large central microprocessor with high-level language capabilities, connected to smaller microprocessors on a common bus, provides an implementation method for the hierarchical control. The software strategy may then be contained in the master controller, and high-speed actions are controlled by the distributed microprocessors.

The distributed sensor and control system is similar in many ways to the central nervous system in the human. Many actions are controlled by neural networks in the spinal cord without conscious control. These local reflexes and autonomous functions are vital to human survival and are important in discovering how they may be simulated in a robot. The study of these mechanisms in robots may ultimately lead to a greater understanding of how we function as humans.

Questions

1. Discuss the concept of common sense and how it relates to humans and robots. Is common sense desirable? When and why?

2. Discuss the question of liability for the actions and possible consequences arising from the use of mobile robots.

3. Compare the use of visual and tactile sensors for a variety of tasks, such as recognizing an object by its shape, color, material, or texture.

Applications of Industrial Robots

6.1 Basic Industries Involved in Manufacturing

The purpose of this chapter is to provide an introduction to manufacturing processes and describe some of the applications of robots in accomplishing these tasks. We will concentrate on proven applications of industrial robots to provide a basis for understanding how they have been used and how they can be used more widely in industry. We will see how robots may be used in all the manufacturing industries. However, to provide a reference point, let's first review the basic processes involved in manufacturing.

Manufacturing is the process by which goods are made available to us as consumers. It involves many industries. These industries concentrate on production. Production is the basic process of transforming raw materials into goods that have a value in the marketplace. The individual steps required are called production operations. Each step adds a certain value to the final product. A look at these basic industries will lead us to an understanding of the scope of applications in which robots are being used and may be used to increase productivity.

The basic industries are represented by public corporations whose stocks are traded on the major stock exchanges. These may be grouped into the following markets (Groover, 1982).

Advertising
Aerospace
Automotive
Beverages
Cement
Chemicals
Clothing
Construction

139

Drugs, soaps, cosmetics
Equipment and machinery
Financial (banks, investment and loan companies)
Foods (canned, dairy, meats, others)
Hospital supplies
Hotel/motel
Insurance
Metals
Natural resources
Paper
Publishing
Radio, television, motion pictures
Restaurant
Retail (food, department stores, others)
Shipbuilding
Textiles
Tire and rubber
Tobacco
Transportation (railroad, airline, trucking, others)
Utilities (electric power, natural gas, telephone)

The companies may also be grouped in several other ways. For our purpose let's first think of a division into companies that produce goods and those that produce services. Although great potentials exist for service-performing robots, we will concentrate on those used to produce goods. The following is Groover's list of goods-producing industries with representative companies listed.

These industries may also be divided into two groups: manufacturing industries and process industries. The manufacturing companies are identified with discrete items, such as cars, computers, machine tools, and the components for these items. The process industries are represented by the food products, chemicals, plastics, soaps, steel, and cement industries.

Robots have mainly been used in the manufacturing industries, as shown in the following table (Robotics, 1983, p. 165).

The RIA estimated that, by the end of 1982, 6300 industrial robots were in use in the United States. Of these, 2453 were used for welding, 1060 for machine loading and unloading, 875 in casting, 1300 in materials handling, 490 in painting and finishing, and the remaining 122 in assembly and other areas. Since the market is new, we may expect some of the distribution to change each year.

Another way to classify the industries is into the three categories of basic producer, converter, and fabricator. Together, then, these three types of industries form a chain in the transformation of raw materials into consumer products, from natural resource to basic producer to converter to fabricator to consumer.

The basic producers take the natural resources and transform these into raw materials used by the other firms. The converter takes the raw materials, such as steel

Basic Industries Involved in Manufacturing

Basic industry	Representative industry
Aerospace	Boeing
Automotive	General Motors
Beverages	Coca-Cola
Building materials	U.S. Gypsum
Cement	Lone Star Enterprises
Chemicals	duPont
Clothing	Hanes
Drugs, soaps, cosmetics	Procter & Gamble
Equipment and machinery	
Agricultural	Deere
Construction	Caterpillar Tractor
Electrical	General Electric
Electronics	Hewlett-Packard
Household appliances	Maytag
Industrial	Ingersoll-Rand
Machine tools	Cincinnati Milacron
Office equipment, computers	IBM
Railroad equipment	Pullman
Steam generating	Combustion Engineering
Foods	
Canned foods	Green Giant
Dairy products	Borden
Meats	Oscar Mayer
Packaged foods	General Mills
Hospital supplies	American Hospital Supply
Metals	
Aluminum	Alcoa
Copper	Kennecott
Steel	U.S. Steel
Natural resources	
Coal	Pittston
Forest	Georgia-Pacific
Oil	Exxon
Paper	Kimberly Clark
Textiles	Burlington Industries
Tire and rubber	Goodyear

ingots, and performs the intermediate step of transforming them into industrial products and some consumer goods. For example, chemical firms convert petroleum products into plastics. The typical output of a converter is in a simple form. The third category of manufacturing firms is the fabricator. These companies assemble and fabricate final consumer products. The final products are either consumer goods or components for consumer goods. For example, automobiles and tires could be final products.

Applications of Industrial Robots

Application	United States (%)	Japan (%)
Welding	35	15
Machine loading	20	40
Foundry	15	
Painting	15	
Assembly	10	30
Other	5	15

Let us now restrict attention to manufacturing rather than process firms, and look at the fabrication firm in particular. The sequence of events between the reception of raw materials or components and the shipping of the final product is called the manufacturing cycle. Typical steps in this cycle are

1. Sales and marketing
2. Product design and engineering
3. Manufacturing engineering
4. Industrial engineering
5. Production planning and control
6. Manufacturing
7. Quality control
8. Shipping and inventory control

These steps vary with the type of industry, kind of product, company size, and management style; however, this division of responsibilities is traditional.

The manufacturing process can be summarized as consisting of four functions:

1. Materials processing and assembly
2. Materials handling and storage
3. Control—from the plant to the operations levels
4. Information system with a manufacturing data base to support the other activities

Materials processing includes those operations that transform a workpiece from one state of completion to another. Basic operations, such as metal casting and plastic molding, would be one step. Secondary processes, such as machining or pressworking, would be the next. Operations to enhance physical properties, such as heat treatment to strengthen metal parts, would be another. Finally, finishing operations, such as painting, polishing, or chrome plating, would be a final process on the workpiece.

Assembly and joining operations are the second type of manufacturing operation. Processes for fastening, such as using screws, nuts, and adhesives, welding, or soldering, might be used. In general, the assembly operations follow the processing operations.

Materials handling and storage is required to get the materials from one operation to another. A significant portion of the total time that a workpiece spends in the factory may be involved in this operation.

The control function can be divided into several levels. At the plant level, the control is concerned with the effective use of labor, proper utilization of resources, shipping products of good quality on time, and keeping plant operating costs at a minimum. At the process level, control involves achievement of performance objectives.

The information function is required to efficiently organize the production functions. A manufacturing data base, including material specifications, part drawings, bills of material, route sheets, tool inventory records, method description, production schedules, and inventory records, must be generated, maintained, and disseminated to support the plant operations.

Industrial robots have been used directly for the materials processing, assembly, and materials handling and storage operations. They may also be integrated into the overall control function of the plant and may directly call upon the information data base. Before considering the fundamental robot applications, let's briefly review the overall strategies that lead to efficient and effective manufacturing.

Groover describes certain fundamental strategies, which he calls automation strategies, that can be used to improve production. These are

1. Specialization of operations
2. Combined operations
3. Simultaneous operations
4. Integrated operations
5. Reduce setup times
6. Improved materials handling
7. Process control and optimization
8. Computerized manufacturing data base
9. Computerized manufacturing control

The realization of these automation strategies is called data-driven automation. The totally automated factory may require the use of all these strategies to provide a competitive edge in the worldwide market. Not surprisingly, manufacturing cost is the largest single factor in determining the selling price of a manufactured product. A typical breakdown (Kutcher, 1983) is

Manufacturing cost—40 percent
Marketing, sales, and administrative costs—25 percent
Profit—20 percent
Engineering costs—15 percent

A graph of the total costs as well as a finer breakdown of the manufacturing costs of a product is shown in Figure 6–1 and listed as

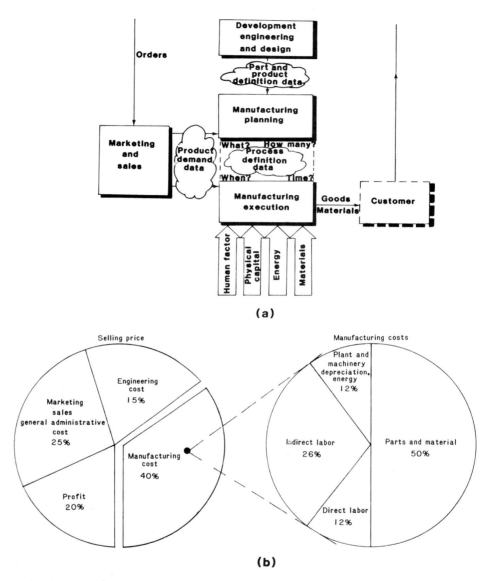

Figure 6–1. Production process and costs of manufacturing a product. (a) Data are both transferred and transformed in a manufacturing enterprise as they go from one entity to another. Product demand data and parts and product definition data are transferred to the planning function in manufacturing. It, in turn, transforms these data into process definition data, which tells the executive function how to produce the product. (b) Manufacturing costs is the largest single factor covering the selling price of a product in batch manufacturing. Direct labor is usually the target of automation but accounts for only 12 percent of manufacturing costs, even though many view it as the sole factor in increasing productivity. Data-driven automation, on the other hand, can dramatically cut all the costs shown. (Adapted from M. P. Groover, "Fundamental Operations," *IEEE Spectrum,* May 1983. Reprinted by permission, Copyright 1980 by IEEE.)

Parts and material—50 percent
Indirect labor—26 percent
Plant and machinery costs—12 percent
Direct labor—12 percent

Improving productivity and maintaining profitability can be accomplished by reducing costs in each of these categories: reducing waste, such as eliminating scrap material, increasing efficiency by automating all areas of the enterprise—white collar, as well as blue collar labor—and minimizing energy and maintenance expenses of the plant and machinery. The integration of computer-aided manufacturing (ICAM) is aimed at improving operations in all levels within the factory.

A research study group was recently established at the National Bureau of Standards to address the problems in manufacturing (McLean et al., 1983). Integration from the facility level to the shop level, the work cell, the work station, and finally to the equipment level is being considered. Each level has its own set of computer controllers for internal use as well as a connection to an overall communications network. Figure 6–2 illustrates the model of the National Bureau of Standards Automated Manufacturing

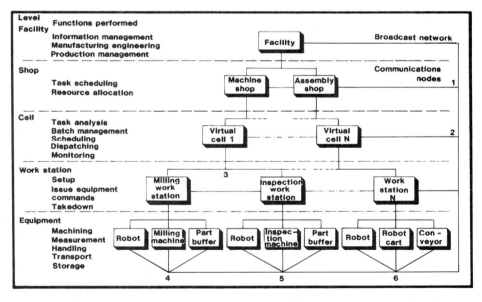

Figure 6–2. Block diagram of the National Bureau of Standards Automated Manufacturing Research Facility. The facility is divided into a five-level command hierarchy: facility, shop, cell, work station, and equipment. Each function box, be it machine shop, milling work station, or robot, has its own set of controllers for its internal control process. All the function boxes communicate along a facility broadcast system through communication nodes. (Adapted from C. McLean, M. Mitchell, and E. Barkmeyer, "A Computer Architecture for Small Batch Manufacturing," *IEEE Spectrum,* May 1983. Reprinted by permission, copyright by IEEE.)

Research Facility. Flexible manufacturing cells with robots are being developed. The entire process is coordinated by computer controllers and a data-administration system.

The overall tasks are divided so that individual work stations receive the parts or materials needed to produce the required product. The equipment included in a work station are the machinery and robots designed to perform the specialized tasks. Let's now consider the types of processing in more detail.

6.2 Fundamental Operations in Material Processing and Assembly

The basic manufacturing operations include machining, forming operations, such as founding and heat treatment, joining operations, such as spot and arc welding, assembly, inspection, and materials handling and storage. Industrial robots have been and are being integrated into the factory of the future in all these operations.

Machining

The machining process is at the very core of manufacturing. The four basic machining processes are turning, drilling, milling, and shaping. When used together, almost any contour can be produced on a workpiece. The turning operation can produce a cylindrical shape on the outside of a workpiece. Drilling can produce a cylindrical space inside the workpiece. Milling and shaping tools create planar surfaces. Modern machine tools are capable of accuracies measured in ten-thousandths of an inch. The speed, feed, and depth of cut determine the time required to produce a part. These three controls determine the rate at which materials are removed from the workpiece as chips or swarf. Figure 6–3a illustrates the four basic machining processes, and Figure 6–3b shows the three variables of speed, feed, and depth of cut for turning on a lathe. This rate is important, since it not only determines the productivity rate, but also the amount of tool wear. The depth of the cut is usually defined by the geometry of the workpiece. Therefore, the speed and feed are the primary control variables. The faster material is removed, the faster the tool wears. Since the tool bears a cost, and replacement takes time, a balance between material removal rate and tool wear must be maintained. High-performance cutting materials, such as tungsten carbide, are now used for making machine tools, and cutting rates of about 2500 feet/minute may be obtained for machine-soft metals.

Another set of operations called forming bends, squeezes, or stretches metal to impart new sizes or shapes, or both. In a forming operation, stresses are applied to deform the part plastically. Sheet metal bending is one example in which a punch and die bend the metal workpiece. Drawing is another forming operation that can be used to change the cross-sectional area of a workpiece by using a pulling motion. This process may be used for wire drawing or metal spinning. Extrusion is the process in which compression forces are used to change the cross section of a workpiece by squeezing it. Cold rolling also uses compressive forces to shape the workpiece. In hot forming operations, the workpiece is heated above its recrystallization temperature but below its melting point. The recrystallization temperature is that point above which the metal will

Fundamental Operations in Material Processing and Assembly | 147

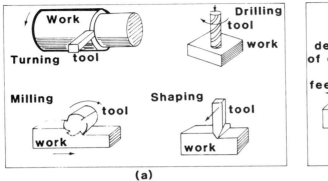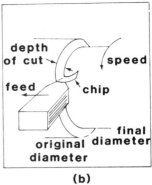

(a) (b)

Figure 6–3. Basic machining processes. (a) The four basic machining processes: turning, milling, drilling, and shaping. A combination of these processes can theoretically produce any desired shape on a workpiece. (b) In any of the basic machining processes, speed, feed, and depth of cut determine the productivity rate. The three variables are shown here for a turning operation on a lathe. (Adapted from M. P. Groover, "Fundamental Operations," *IEEE Spectrum,* May 1983. Reprinted by permission, copyright by IEEE.)

form larger, strain-free grains when it re-forms. Heat forming is usually employed with compressive processes. Cold forming is done below the recrystallization temperature to increase the strength of the material.

Presses are used for many metal forming operations. Shearing, which deforms a material past its breaking point, may be used to cut one portion of the material from another. Embossing may also be done on a press. Figure 6–4 illustrates some of these forming operations.

Some of the newer machining methods that have come into use are electric discharge, ultrasonics, chemical, and laser beam machining, and plasma-arc cutting. Electric discharge machining uses a highly concentrated, high-energy electric spark to remove metal. Ultrasonic machines direct abrasive particles in a liquid slurry against the workpiece surface. In chemical machining, corrosive substances that dissolve metal in a controlled manner are used. Laser beam machining uses a high-energy laser beam to make small, precise cuts. Finally, in plasma-arc cutting, high-temperature directed plasma beams are used to cut sheet and plate metal.

Robots may be used to perform these operations by moving a process tool, such as a laser, around the workpiece, or by moving the workpiece under the process tool. Let's now examine some robot applications in the materials processing area.

Processing of Raw Materials

The processing of raw materials into a form suitable for other manufacturing processes is the beginning of the production cycle. The melting and heat treatment of metals provide an excellent example of these processes. The melting and pouring of a substance, such as metal, into molds to set in the shape of the mold is called the "founding" process.

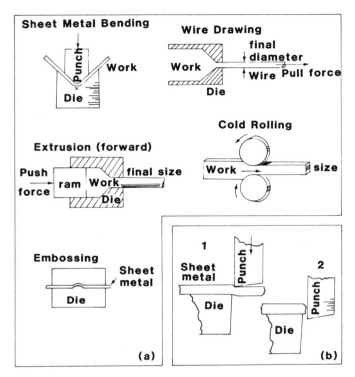

Figure 6–4. Basic forming operations. (a) Sheet metal bending, wire drawing, extrusion, cold rolling, and embossing are forming operations that bend, squeeze, or stretch metal to impart new sizes or shapes. (b) Shearing produced by a punch and die deforms metal beyond its breaking point, thereby separating one portion of metal sheet from another. (Adapted from M. P. Groover, "Fundamental Operations," *IEEE Spectrum,* May 1983. Reprinted by permission, copyright by IEEE.)

Founding is a primary manufacturing process. However, in terms of work environment, human operators may be required to work in high-temperature, noisy environments with noxious fumes, splashing molten metal, and large pieces of moving machinery that can be hazardous. Yet this is at the very start of the manufacturing cycle, and thus all later steps are dependent on it. Automobiles, farm machinery, and many other industries depend upon the materials supplied by foundries. Founding is ranked sixth in the United States on the basis of value added to the raw materials by the processes performed on it (Engleberger, 1980, p. 225). Robots are used to load and unload castings, presses, and many other machines for materials processing tasks.

Casting Process

The casting process consists of four main steps. First, the material is heated until it is molten. Then, it is poured into a mold. After the material has cooled, the mold is removed. Finally, the casting is finished. Finishing may involve removal of excess material or other operations.

Robots have been used in foundry applications for preparing the molds, ladling the materials into the molds, removing the castings, and for finishing operations, such as deflashing (Engelberger, 1980, Chap. 18).

Heat Treatment

Materials are often processed by heat treatment operations designed to modify the atomic structure to improve such characteristics as hardness, strength, ductility, or electrical conductivity. The process consists of placing the material in a furnace where it undergoes controlled heating, then removing the material and placing it in a location for controlled cooling.

Such parts as forgings, castings, or cold worked materials are often heat treated. The type of operation required may be a simple pick-and-place operation, which is easy for a robot. Furthermore, because of the high temperatures encountered, the work environment is more suited to robot rather than to human operation.

Robots are ideal for working in such high-temperature environments. One application is shown in Figure 6–5. In this example, which is from an International Harvester factory, harrow disks are heat treated to toughen the parts against breakage when struck by objects, such as rocks, when in use. In this operation three robots are used. Robot 1 stands at the entry conveyor system for the furnace. Its hand is equipped with a vacuum gripper. The gripper comes down vertically and lifts the top disk from a palletized stack of about 50 disks and transfers the disk to the conveyor, which carries it into the furnace. The furnace temperature is 1650 degrees Fahrenheit. If the robot hand finds no disk on the stack, the vacuum cup rests on the framework of the pallet, and a pneumatic pressure sensor terminates the program while a new stack of disks is moved into position. When this is accomplished, a signal restarts the program.

Robot 2 is equipped with a two-fingered gripper that grasps the disk on its outside diameter. As each disk emerges from the furnace it continues on the conveyor until it reaches a "pop-up" station, which positions it for the robot to pick up. A gating system holds subsequent disks until the pop-up station has dropped back into position. Robot 2 then picks up the disk, rotates 160 degrees, and loads it into the die of a press. Before the robot hand enters the press, three conditions must be met. First, the die must be open, which is indicated by a ram limit switch. Next, robot 3 must signal that it has removed the previous disk from the press. Finally, an infrared scanner must indicate that the disk has reached the proper operating range of 450 to 600 degrees Fahrenheit. If the sensor indicates that the disk is too hot or cold, robot 2 switches to a reject program. Robot 3 unloads the press and swings around to place the disk on the incoming conveyor of a washing and drying unit, which cleans off the solid salt remaining from the quench process.

This example clearly illustrates the effectiveness of robots in heat treatment processing.

150 | Applications of Industrial Robots

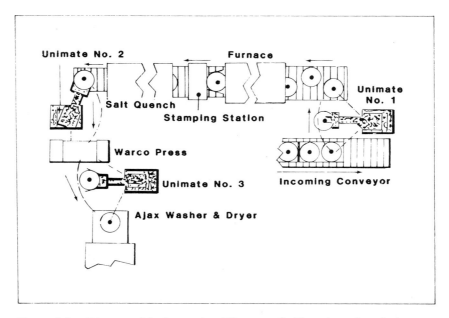

Figure 6–5. Diagram of the International Harvester facility using robots for heat treatment of harrow disks. (Courtesy of Joseph F. Engelberger.)

Welding Applications

Spot Welding. Welding is the most common robot application in the United States. Welding is the process of joining metallic parts by heating and allowing the metal to flow together or fuse. In spot welding, metal parts are joined at a number of small localized areas. This is accomplished by passing a large electric current at low voltage through the metals, which are held together under high pressure. The electric current generates heat from the work required to overcome the small but finite resistance of the metals being joined. In practice, the metals are clamped together with a high pressure between copper or copper-alloy electrodes, which conduct the welding current to the weld spot. As current flows from the power source, through the electrodes and through the workpiece, heat is generated at the point of contact. If the heat is sufficient to melt the materials, a fusion of the materials takes place. The amount of current and duration of flow must be controlled to provide high-quality welds. Poor-quality welds may have no fusion or may burn through. Modern spot-welding machines are equipped with automatic controllers that can be set up to perform the required sequence of weld operations. The sequence consists of a squeeze step in which the two electrodes are forced together with a pressure from 800 to 1000 pounds per square inch. Next, the weld step in which the current is turned on and flows through the materials is executed. Then, a hold step is performed in which the current is turned off but the tips are held together long enough for the materials to cool. Finally, a wait step is included in which the machine is turned off until the tips are cool enough for the next operation.

Fundamental Operations in Material Processing and Assembly | 151

Spot welding is most suitable for ferrous metals, which are electrical conductors with enough resistance to generate the desired heat. Applications for spot welding include automobile bodies, appliance cases, and sheet metal fabrication. The spot-welding guns may weigh as much as 200 pounds since they include the heavy-duty movable electrodes, electrical cables capable of carrying as much as 1500 amperes of current, and often a coolant, such as water, for the electrodes. Spot welding requires positioning the welding gun perpendicular to the workpiece. This requires great dexterity but is ideally suited for a 6-degree-of-freedom industrial robot.

Robots for spot welding are widely used in the automotive industry. One example of a spot-welding system is shown in Figure 6–6. This application was developed for the Chrysler Newark assembly plant and uses a shuttle system that moves the automobile bodies off the main assembly line for the spot-welding operation. The shuttle line has seven stations with a total of 12 Unimate robots. The line is indexed so that the welding operations may be performed on stationary workpieces. Each robot performs a sequence of welds at its station. After the welding is completed the workpiece is transferred back to the moving line. One of the main advantages of robot welding in this application is in the consistency of the welds. Consistency permits a reduction of the number of weld locations in comparison with those required for human welds.

Another example of a spot-welding application is shown in Figure 6–7, which clearly shows the hazardous sparks generated in a spot-welding application.

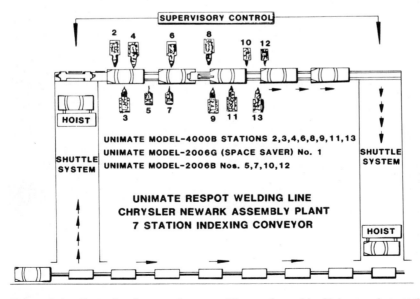

Figure 6–6. Example of spot and seam welding performed by Unimate robots at the Chrysler Newark assembly plant. A shuttle conveyor system is used to provide a stationary workpiece during the spot-welding operation and not slow down the assembly line. (Courtesy of Joseph F. Engelberger.)

152 | Applications of Industrial Robots

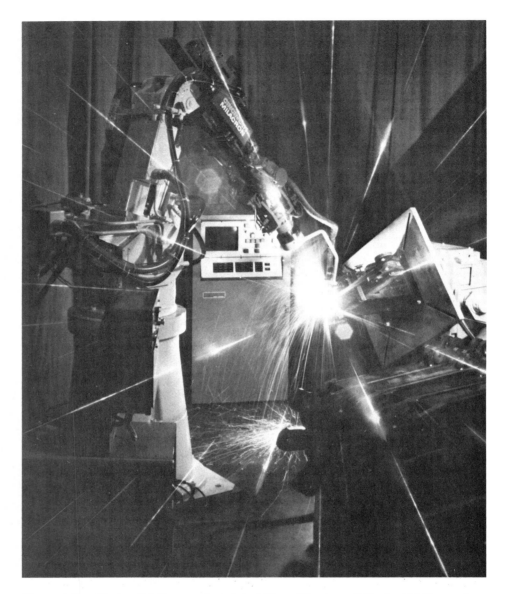

Figure 6–7. Cincinnati Milacron robot spot welding. (Courtesy of Cincinnati Milacron.)

Arc Welding. Whenever a gas-tight seal or a long path weld is required, spot welding is not appropriate. The arc-welding process can accomplish these procedures. The electric arc-welding process fuses the metal surfaces together with the heat generated by an electric arc between an electrode and the workpiece. The arc is generated between the two by connecting the workpiece to the power source so that it becomes the second electrode. The electric arc welder generally operates from a direct current source capable of supplying 100 to 200 amperes of current at 10 to 30 volts.

In manual operation, the operator connects the workpiece to one terminal of the supply with the electrode connected to the other. To start the arc, the operator touches the electrode to the workpiece. This essentially short-circuits the supply and starts a large current flow, which is accompanied by heat generation and sparking. The operator then withdraws the electrode a short distance and maintains an arc discharge. If the distance is too great, the arc stops. If the distance is too short, the electrode may stick to the workpiece.

By monitoring the voltage between the electrode and workpiece a signal can be obtained that may be used to regulate the distance. Before contact, this voltage will equal the supply voltage. Upon contact, the voltage drops to near zero. At the correct distance, an intermediate value of voltage is obtained.

For high-quality welds, the material in the weld pool must also be controlled. If the electrode is not made of a material similar to that of the workpiece and melts into the pool, an inconsistent weld may result. One solution is to use electrodes made of tungsten, which has a much higher melting temperature than the typical workpiece. However, if excess material is needed it must be added from a separate filler supply. Also, oxidation reactions with the molten materials must be avoided to prevent inconsistent weld density. One method used to prevent oxidation is to flood the weld pool with a flux or an inert gas, such as helium or argon.

The most commonly used method for arc welding on an assembly line is called metal inert gas (MIG) welding. This process uses a continuously fed metal electrode selected for the particular material to be welded to provide a filler material. Also, an inert gas surrounds the arc and weld pool to shield the pool from any oxidation reaction. The filler and gas are supplied by threading the electrode through a gas line. The control equipment feeds the wire at a selected rate and regulates the flow of the gas.

Proper arc welding requires accurate location of the welding gun along the weld path in position, orientation, and speed. A typical sequence of operations to perform a weld starts with a preflow condition in which the gas is turned on. Next, the weld period commences in which the wire feed is started and the power applied. After the weld is finished, a burn-back period is used in which the wire feed stops and the wire tip burns off until the distance is too great to sustain the arc. Next, a postflow period is started in which the power goes off with the gas flow continuing, permitting the weld to cool. Finally, the sequence is completed and the gas flow stops.

A Unimate apprentice robot is shown in Figure 6–8. Note that the operator, who is simply supervising the robot, must wear protective goggles to avoid the infrared radiation emitted by the arc and protective clothing to prevent burns. Also, since an ozone atmosphere is generated that cannot be breathed, using the robot permits the operator to work at a safe distance.

An ASEA robot is shown arc welding in Figure 6–9. Note that the operator is not in the hazardous environment. Next, a Hitachi welding robot is shown in Figure 6–10. This robot is one of a pair used to assemble, tack weld, and seam weld motor frames. The frame is a steel tube, which is moved into position in front of the robots by conveyors. One robot picks up the cooling fins, support braces, and conduit box mounting support

Figure 6–8. Unimate apprentice robot. (Courtesy of Joseph F. Engleberger.)

Figure 6–9. ASEA welding robot. (Courtesy of ASEA.)

Fundamental Operations in Material Processing and Assembly | 155

Figure 6–10. Hitachi robot seam welding. Over 500 hardworking Hitachi Process Robots have been installed in industrial applications, many of which have been arc welding. This photograph demonstrates a very good example of flexible automation. Shown is one of two Process Robots working as a team to assemble, tack weld, and then seam weld (5 to 200 horsepower) motor frames. A pretaught program for a particular frame size is selected at the touch of a button. The frame is a steel tube, which is moved into position in front of the Process Robots by conveyors. One robot picks up cooling fins, support braces, and the conduit box mounting support and locates the items one at a time on the tube frame. The other robot tack welds and then seam welds each item to the frame. Welds are high quality, which assures good heat transfer. The conveyor then removes the completed motor frame and brings the next tube. (Courtesy of Hitachi America, Ltd., Allendale, New Jersey.)

and locates the items one at a time on the tube frame. The robot shown tack welds and then seam welds each item to the frame. The completed motor frame is then placed on the outgoing conveyor. A Cincinnati Milacron robot is shown welding a curved metal flange of a pipe in Figure 6–11. The dexterity of the three-roll wrist and the controlled path motion are especially important in welding curved surfaces.

Assembly Operations

Discrete parts assembly is a growing application area for robots in the United States and is the largest application in Japan. Typical assembly operations include picking, placing, fastening, connecting, and using such tools as screwdrivers and wrenches. Assembly operations require greater repeatability and more sensors than most other applications. However, the advantages of consistent quality and improved production rates provide motivation for many applications.

The Cincinnati Milacron T3 726 robot is shown inserting screws in an assembly

156 | Applications of Industrial Robots

Figure 6–11. A Cincinnati Milacron welding robot. (Courtesy of Cincinnati Milacron.)

operation in Figure 6–12. The robot shown could work side by side with humans on an assembly line; however, humans could not maintain the speed of operation for this task. A General Electric robot is shown in a large part assembly application in Figure 6–13. In this application, the GP 132 robot is shown assembling the two major components of a Hotpoint refrigerator in the GE Cicero manufacturing plant. The parts are fed from two conveyor lines, and the GP 132 is mounted on a traversing base so that it can pick up moving parts and move between fixtures. In this application, the robot replaced the lifting task required by one person who removed liners from an overhead conveyor and two who placed the liner in the case.

Finishing Applications

Many finishing operations are also performed by robots. Sharp edges on objects are often produced in machining operations. A deburring operation is used to shape the edges of the workpiece to a given contour. This deburring operation is illustrated in Figure 6–14, which shows a Cincinnati Milacron T3 726 robot maneuvering an air-powered deburring tool around the complex contour of a housing. Polishing and buffing operations may also be accomplished.

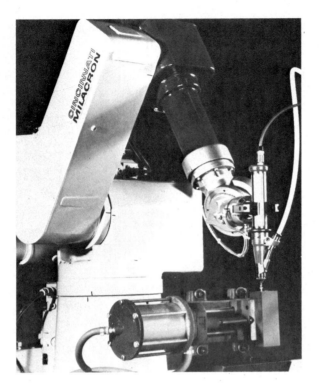

Figure 6–12. A Cincinnati Milacron robot using a screwdriver in an assembly operation. (Courtesy of Cincinnati Milacron.)

Spray painting is currently the largest finishing application for industrial robots. The spray-painting process is used to coat a surface with a liquid mixture, with a solid pigment suspended in the liquid. Painting may be used for protection or decoration. The technique and special fast-drying paints were developed in the automotive industry to provide a high-quality finish for this important mass-produced item.

In manual spray painting, the operator holds a pressurized spray gun, which is fed from a paint reservoir. The distance from the gun to the workpiece is critical, to provide an even coating and avoid runs. The operator usually moves the gun back and forth over the surface to apply thin coatings that build up to a layer of even thickness.

Spray-painting robots not only provide greater productivity by increasing throughput, but also reduce paint costs by eliminating wasted paint, reduce energy requirements by eliminating needless motion, reduce rejects by providing consistent quality, and increase production flexibility because of robot programmability.

Programming a spray-painting robot is often said to be done best by the best painter on the best day, as already mentioned. Continuous path motion is used to provide the control path for the robot. It is difficult for a human to repeat a continuous motion path.

Figure 6–13. General Electric Company's GP 132 assembly robot. In this application, the GP 132 is assembling the two major sheet metal components of a Hotpoint refrigerator in the GE Cicero manufacturing facility. The parts are fed from two conveyor lines, and the GP 132 is mounted on a traversing base so it can pick up moving parts and move between fixtures. The two parts to be assembled are the liner and case. The liner is the white sheet metal part fed from an overhead conveyor and suspended on a swiveling hanger. The robot body moves along its traversing base to match the speed of the part on the conveyor as it removes the part from the hanger. An array of six vacuum suction cups grabs the liner and allows the robot to move it. After briefly stopping to align the liner in a stationary fixture, the robot then places it inside the case. The case is lifted above its conveyor line by hard automation (piston and cylinder and others) and held stationary during the insertion procedure. Inserting the liner is a delicate procedure since any side-to-side motion of the liner in the case would result in an unacceptable deformation or breakage of the Styrofoam seal on the inside edge of the case. The robot replaces one person who removed liners from the overhead conveyor and two who placed the liner in the case. The GP 132 is a six-axis, electric robot designed for loading and unloading, spot welding, palletizing, and other materials-handling applications. It can handle up to 132 pounds. (Courtesy of General Electric Co., Bridgeport, Connecticut.)

Fundamental Operations in Material Processing and Assembly | **159**

Figure 6–14. A Cincinnati Milacron robot used in a deburring operation. (Courtesy of Cincinnati Milacron.)

Automatic part identification and color changing may also be used in robotic spray painting, which eliminates the need for a human to be in the spray environment.

A typical layout of a spray-painting booth is shown in Figure 6–15. The robots may be enclosed in a booth with a controlled air supply in a compact arrangement that provides efficient use of floor space and protects humans from a hazardous environment. An important feature of the industrial robot in this application is its programmability. A continuous throughput can be maintained for a particular part once the operation is set up. However, even when the workpiece changes, only the program requires changing,

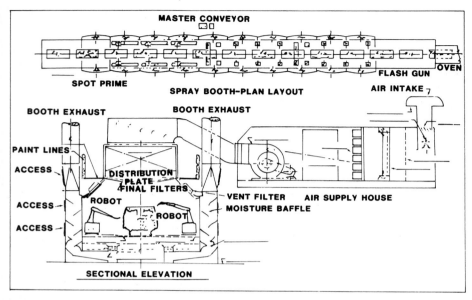

Figure 6–15. A spray-painting booth layout. (Courtesy of Joseph F. Engelberger.)

which is much easier than changing the major machines. Furthermore, much of the programming can be done off-line before the change and simply downloaded to the robot controller.

A DeVilbiss/Trallfa robot is shown in Figure 6–16, painting metal shutters in a finishing operation. The various positions of the spray gun are shown in Figure 6–17, in which a Nordson Company robot is stimulating painting the interior of an automobile.

6.3 Materials Handling and Storage Applications

Industrial robots are excellent for performing many materials handling and storage applications, such as machine loading and unloading, material transfer, palletizing, and bin picking. In many applications, back-breaking tasks for humans are easily done by a robot.

Machine Loading and Unloading

The general layout of a flexible manufacturing cell is shown in Figure 6–18. The robot is placed in the center with the machine tools within its work space. A cycle time analysis with various machine placements may be used to optimize the placements. Several machines are grouped around the robot because its operations are much faster than the operations of the machine tools. This clustering of machine tools around a central robot

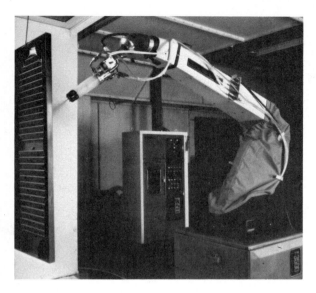

Figure 6–16. DeVilbiss/Trallfa robot painting metal shutters in a finishing operation. (Courtesy of DeVilbiss Co., Toledo, Ohio.)

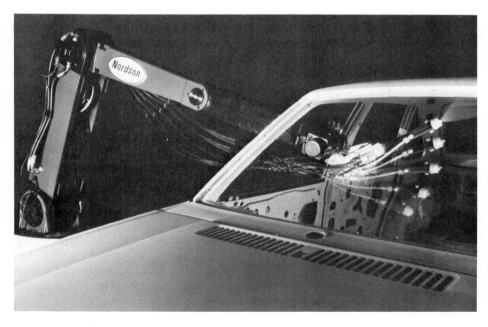

Figure 6–17. Nordson spray-painting robot. (Courtesy of Nordson Co., Amherst, Ohio.)

162 | Applications of Industrial Robots

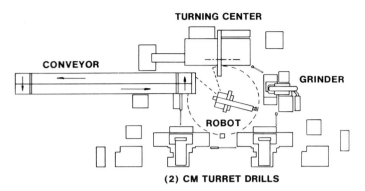

Figure 6–18. A diagram of a flexible manufacturing system. (Courtesy of Cincinnati Milacron.)

provides a flexible manufacturing operation not only because of the flexibility of the robot but also because modern machine tools also have great flexibility, including automatic tool changing. During operation, parts are fed from the incoming conveyor. The robot picks up the part and moves it through the desired cycle. At the completion of the processing, the robot places the finished part on the outgoing conveyor. An actual cell with Cincinnati Milacron robots and machine tools is shown in Figure 6–19.

A Prab Robots, Inc., robot performing machine loading and unloading operations as the central element in another flexible manufacturing cell is shown in Figure 6–20a. The robot is again capable of servicing several different machine tools, including three drilling and boring machines. The installation of this work cell was at the Fluid Power Operations of Eaton Corporation in Marshall, Michigan. The part to be machined was a 20-pound malleable iron casting used for a locking differential for three-quarter-ton trucks. A pair of bell-shaped housings arrive at the work cell on an indexing conveyor precisely oriented on a fixtured pallet for the robot to pick up. Photoelectric cells are used to communicate to the robot that parts are in position and ready to pick up. The machines are designed to pick up two housings at a time. To accomplish this, the robot's gripper was designed to pick up and simultaneously move two parts through the three machines.

The first machine performs a drilling operation. This is followed by a boring operation in the second machine. Both these machines require the parts to be oriented end to end with the housing sides exposed for machining. The third machine performs a drilling operation and requires the parts to be oriented side by side. The dual grippers mounted on a rotary cylinder attached to the robot arm accomplish this repositioning.

Following the last drilling operation, the robot moves the parts to an unloading fixture on an outgoing conveyor. The actual installation is also shown in Figure 6–20b. During the first 2 years of operation, productivity was increased by 60 percent, direct labor costs were reduced, and a back-breaking job was eliminated.

This example is typical of the many machine loading and unloading applications of the industrial robot. The robot is the central element of the flexible manufacturing cell.

Materials Handling and Storage Applications | **163**

Figure 6–19. An actual robot manufacturing cell consisting of two CINTURN NC turning centers, gaging station, incoming stop-station conveyor, and unload system. (Courtesy of Cincinnati Milacron.)

Palletizing

Palletizing is another important robot application. If the objects to be palletized are of the same type, an indexing program may be easily developed. An example is shown in Figure 6–21. This 19-foot-tall palletizing robot is capable of lifting 1300 pounds, which far exceeds the capability of a human. The robot shown is lifting heavy rolls of plastic.

Another palletizing example is shown in Figure 6–22. A vacuum gripper is used to lift containers and place them on a pallet. This heavy lifting task is easily accomplished by the heavy-duty Cincinnati Milacron T3 industrial robot. The programming for palletizing is also simplified by a special indexing feature in the T3 language. Palletizing fixed-size parcels, which are normally encountered in manufacturing, is a well-proven application of industrial robots.

Interestingly, if the parcels are of mixed size, shape, and weight, as may be encountered in the distribution stage, the problem is much more complicated. Sensors to determine these sizes and shapes are required. Also, an artificial intelligence algorithm for filling a three-dimensional space with these parcels is needed. The general form of

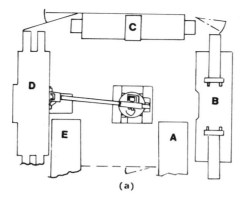

Figure 6–20. (a) Line drawing of a machine loading and unloading work cell. (b) Machine loading and unloading by a Prab Model FA robot. The work cell is installed at the Fluid Power Operations of Eaton Corporation in Marshall, Michigan. The part to be machined is a 20-pound malleable iron casting used for a locking differential for three-quarter-ton trucks. A pair of bell-shaped housings arrives at the work cell on an indexing conveyor precisely oriented on a fixtured pallet for the robot to pick up. Photoelectric cells communicate to the robot that the parts are in position and ready to pick up. The machines are designed to accept two housings at one time. Therefore, the robot's end effector is designed to pick up and move two parts simultaneously through the three machines. The first machine performs a drilling operation. This is followed by a boring operation by the second machine. The first two machines require identical positioning of the part, but the third machine performs a drilling operation requiring the parts to be oriented side by side. Between the second and third machines the parts are repositioned. The dual grippers mounted on a rotary cylinder attached to the robot arm are programmed to accomplish this repositioning. Following the last drilling operation, the robot moves the parts to an unloading fixture on an outgoing conveyor. The robot performs a sequence of 49 steps to complete the cycle. (Courtesy of Prab Robots, Inc., Kalamazoo, Michigan.)

Materials Handling and Storage Applications | **165**

Figure 6–21. A materials handling robot. The Positech Corporation's Probot is a heavy-duty palletizing robot that can lift up to 1300 pounds and is 19 feet high. The robot has 100 inches of reach, 100 inches of lift, and 350 degrees of rotation for its cyclindrical coordinate design. In this picture, it is shown lifting plastic rolls at a factory in Belgium. (Courtesy of Positech Corp., Laurens, Iowa.)

this problem with arbitrary shapes to be fitted into a space is still unsolved mathematically. Interesting games are often made that are variations of the problem with a limited number of different sizes and shapes.

Bin Picking

Another interesting research problem in a materials handling application is called bin picking, which consists of selecting a single object from a bin of objects. Several interesting solutions to this problem have been developed by Professor Robert Kelley at the University of Rhode Island. One solution consists of using a special gripper, which

Figure 6–22. A Cincinnati Milacron HT3 robot stacking boxes in a palletizing operation. A special software function called index makes such operations easy to program. (Courtesy of Cincinnati Milacron.)

can pick up a single part and then move the part in front of a camera for recognition or inspection. A commercial realization is now being developed.

Questions

1. Select a simple manufacturing task of interest, and analyze the automation of the solution using a robotic work cell.
2. Visit a nearby manufacturing facility. Note the current and possible applications of robots.
3. Consider the process of manufacturing paper clips, starting with the raw ore and ending with the final delivery of cases to a distribution warehouse. Would robots be useful in this process? Where?

Economic Considerations and Justification of Industrial Robots

7.1 Economic Considerations

When robots were first developed, they were very expensive, requiring large amounts of capital investment in a new, untried technology. Now, however, many robots have been installed with great success, leading robot producers to manufacture more robots at lower prices. Although robots are still a considerable investment, they no longer represent an untested production method. Thus, more businesses are deciding to invest in robots. The purpose of this chapter is to consider the economic aspects of robot ownership by industry or by individuals.

The cost of an industrial robot is often less than that of a house. The price range for industrial-quality robots in the United States is between $14,000 and $150,000, according to the RIA *Worldwide Directory*. A robot is therefore within the purchasing power of individuals or groups of individuals. What circumstances would lead one to purchase a robot? When can the purchase of one or more robots be justified? What conditions must be met to make these purchases attractive to the investor? These are some of the questions we consider in this chapter.

The prices of robots run from about $35 for a Tomy toy manipulator arm to $100 million for the space shuttle manipulator arm. Computer-controlled robots start at about $300 for the Turtle. High-accuracy industrial robots are around $50,000. This range of prices shows that robots are in the range of affordability for many people, but what would we have it do? Hugh Hefner has one to bring him drinks and greet guests. Our son told us he would like a robot to make money for him so that he could buy all the toys he wanted. Then he would want the robot to help him bring all those toys home and help him put together any that came unassembled. We all would undoubtedly like a robot to make money for us so we could buy what we want. What can a robot do to make money? Well, it can work, that is, produce goods or provide services, just like humans. We have already discussed several of the jobs that robots can do—welding, painting, parts manipulation, and assembly. What can a robot build?

As an exercise in a robotics class, the students were asked to design a single robot system that could build computers. The motivation for this project was that if a single robot could take the components of a computer and put it together, any student could have a computer for the cost of the components, which is about 40 percent of the usual retail price of most computers. After 3 months, the students had come up with designs of several systems that could assemble a computer. The size of the best overall system was about that of a dining-room table. The cost of the robot and sensors was about $50,000. The time required to assemble the small, single-board computer was about 5 minutes. Running 24 hours/day, this robot assembly system could theoretically produce 288 computers per day, or 105,120 per year. This could supply even the largest college with at least one for each student, faculty, and staff member. If only $100 were saved on each computer, this single robot system could save $1,051,200 in 1 year as compared with purchased systems. A $1 million return on a $50,000 investment is rather attractive. The students did not have time to actually build the systems they designed, but similar exercises are being done in industry for real dollars. Many brilliant designers are probably thinking about how robots can make money, but we may never learn of their efforts until we see their results at the local shopping mall in the form of better products at lower cost.

Industrial ownership of robots is currently the most economically practical method of ownership in our country. Any major departure from our existing free enterprise system is highly unlikely in the near future. Corporations have the resources, the skilled designers, and the artisans to make just about anything we will buy. Competition between domestic and international manufacturers is very tough. As consumers, we benefit from this competition. As competitors, however, we must use every possible resource available to stay in the race. Robots are a very useful resource.

What are some of the considerations on which potential buyers of robots need to concentrate? Probably the most important is how to justify the costs associated with buying and using robots. We perform a sample economic justification and break-even analysis later in the chapter, but to get an overall idea of how the cost of robots compares with the cost of alternative forms of manufacturing, let's examine the basic reasons robots are now being considered by many businesses that formerly believed robots were too expensive.

The need for robots arose from the need for flexible, programmable automated machines to perform a variety of tasks in a variety of situations. The major benefits arising from the use of industrial robots, such as increased productivity, improved product quality, greater resistance to inflation or sudden product changes, fewer employee injuries, continuous output, and better inventory control, made robots desirable to many businesses. However, the cost of robots as compared with human labor costs precluded their purchase except by large or wealthy businesses. Then, the cost of labor began to rise. It is now at a point where the robot has become less expensive in

170 | Economic Considerations and Justification of Industrial Robots

comparison. Figure 7–1 shows a graph of hourly labor costs as compared with robot hourly costs from 1960 to 1981. Note that the ratio of human-robot labor costs is almost 4:1 in this example.

The next most important consideration in determining whether a robot purchase is justified is how robot systems can improve productivity. Robots can do this in several ways. They can improve throughput efficiency, consistency, and product quality. In some applications, such as arc welding, productivity can increase several hundred percent, depending on labor costs, operations, and other related costs. Another obvious benefit in productivity is improved product quality and consistency. This improvement is made because robots, unlike humans, cannot grow tired, bored, or careless with their work. When humans tire, they sometimes make mistakes in their work, resulting in the need to scrap or rework parts of the product. This costly reworking and scrapping of unfit products is nearly eliminated with robots. The robotics system will produce the same part exactly the same way many thousands of times without a breakdown. Robots can also represent a savings in direct labor costs in its particular operation, since one robot can usually do the work of three to five humans. There are also savings in associated costs, such as lighting and heat, that are incurred in a workplace where human comfort is a consideration. Robots can work in the dark and in temperatures uncomfortable for

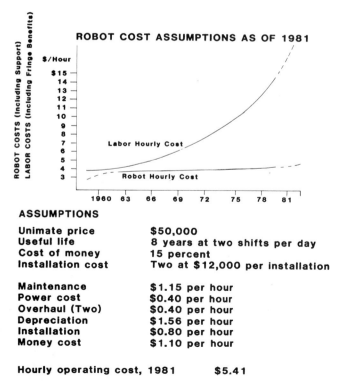

Figure 7–1. History of labor and robot costs. (Cost estimates supplied by Unimation, Inc.)

humans. Savings may also be made in safety costs, such as those resulting from compliance with government standards for high-risk jobs. Savings may result from more accurate forecasting of production schedules. If a robot system is programmed to produce a certain number of parts per hour, then that is how many parts will be produced, barring mechanical failure. Failure rates among the robots produced by major manufacturers are usually quite low. Many robot designs reach a mean time between failure of 500 hours without incurring prohibitive costs.

The next most important consideration is the type of robot hardware, software, and power supply called for. Although we have already described the various types in earlier chapters, let's go over them very generally again. The manipulator of a robot can vary according to the kind of dexterity the potential application requires. The number of manipulator axes that are available on most industrial robots ranges from two to six. The four types of robot designs most widely available are based on their manipulators' geometries. These are Cartesian, cylindrical, spherical, and articulated. Each different application requires an end effector designed to perform the needed task. These can be grippers, drills, magnets, spray guns, welding units, or others. Many tasks sometimes require custom-built end effectors, although standard welding guns, spray-paint guns, and grippers are available for robots. The power supply for the robot can be hydraulic, pneumatic, or electric. Hydraulic-powered robots are generally called for when the application involves handling heavy materials, or where dexterity and resilience are required. Pneumatic robots perform best when the application calls for very rapid movement or in handling very light materials. Electric robots can achieve a higher degree of repeatability and accuracy as long as the payload involved is not heavy. The current trend is toward all-electric robot systems, particularly when the system involves sensory accessories. The controls for the system can range from basic programming controls to minicomputers, depending on the type and complexity of the operation required of the robot system. In most applications calling for frequent human intervention and/or adjustment, such as in arc welding, an operator control station that has a teach pendant, keyboard, and CRT display is supplied.

The next most important consideration is the robot system software. Control software is supplied with the robot control. Applications software to integrate the robot with other equipment in the factory is custom-built. The complexity and expense involved in the software again depend on the application of the robot system. It can range from simple single-cell robot controls to hierarchical computer packages that interface with robot communications. The type of software needed is also affected by the type of robot operation called for in the application, whether servo or nonservo. Servo robots can perform point-to-point, continuous path, and controlled path operations; nonservo robots are limited to simple line transfer and materials-handling operations. Most robots manufactured and sold today are of the servo type.

Another important consideration for the user is to what kind of application robot systems are best suited. Proven applications for robot systems have been detailed in Chapter 6, but let's go over the list of most frequent applications to date. These are:

Spot welding
Arc welding

Palletizing
Stacking and unstacking
Stamping press loading and unloading
Drilling
Painting
Handling and moving hot parts
Assembly
Machine loading and unloading
Injection molding
Die casting
Shell molding for investment castings
Deburring, grinding, and milling
Gluing
Handling and moving toxic, dangerous, or heavy parts

This list is by no means exhaustive, but because they are proven applications, they represent references for the potential buyer and user. Robots can be beneficial in both large and small operations, for both long runs and small batch operations. Robots can come freestanding, in single-cell applications packages, or in integrated applications systems that can be interfaced with other manufacturing equipment. An evaluation of a robot's potential benefits for a particular facility should be made. In making such an evaluation, the following questions should be answered for each individual application.

Is there a sufficient volume of parts or number of operations the robot will perform to make its purchase economically feasible? Or, is the volume so large and unchanging that hard automation is more economical?

Is the contemplated robot system compatible with the present process? That is, are the changes that will be necessitated by the system feasible and cost-effective? Is the production line speed needed in the application within the limits of the contemplated robot system?

Can the robot system be reprogrammed or retooled to accommodate anticipated product or operation changes without prohibitive additional cost?

Is the robot system capable of handling the required payload with maximum efficiency?

Does the contemplated system have sufficient operating range?

What degree of repeatability is required for the application?

Are software and memory available for the contemplated system in sufficient degree to handle anticipated changes or adjustments in the application?

Does the control system need to be easily or frequently changed? How easily can the robot system be retrained?

How much adaptability and interface is required of the sensors?

How much robot can the facility handle? Are there adequate floor space, floor strength, and power supply? Some industrial hydraulic robots can weigh 4 tons.

How much downtime can be accommodated? The industrial standard for most robot systems is currently about 2 percent, but this does not include peripheral equipment. In most applications, the uptime reported for robots is 98 percent. This indicates that the robot would be down only 40 hours in a 2000-hour year.

Stauffer (1982) summarizes the four basic types of robot installation. These are the single, stand-alone robot, multiple cells, each with a robot, an integrated line with similar robots performing similar tasks, and an integrated line with different robots doing various tasks. He reports that, since the current trend is toward integrated installations, one of the most important considerations for the potential user is how future requirements will fit into the robotic system they choose. For example, if there is a possibility of adding a robotic vehicle transport system, automatic storage and retrieval systems, or off-line CAD programming within 10 years of the first robot system installation, then that first system should be chosen carefully to accommodate these anticipated systems.

Computer simulation of robotic systems is a valuable tool that allows potential users to predict production times and to comprehend how the system will work, how the system can be integrated into the present production process, how the selected system compares with alternative production processes or systems, and how anticipated additions, such as additional systems or peripherals, can best be integrated into that system. These simulations can be in color and in three dimensions and can be done by independent organizations to ensure objective comparisons.

Before performing an economic analysis, the human relations aspects should be openly and honestly considered by management, shop employees, union representatives, and any others concerned with the company remaining in business.

The human factors involved in the application of any robot installation are of paramount importance because the system will only be as successful as the people involved allow it to be. Howard (1982) lists the following factors that should be considered before integrating any robotic system in a factory.

1. *Work environment.* Assess the physical and psychological surroundings and the adaptations workers will make in response to them.
2. *Worker/machine interface.* Analyze the manner in which the humans and the machines will function together to design a system that maximizes the capabilities of both. For example, if the robot system will move at one speed and humans at another speed, the production must be coordinated to accommodate both without penalizing either.
3. *Job design.* Analyze the capabilities and limitations of humans and machines to determine which jobs are most satisfactorily and/or efficiently performed by humans and which jobs should be performed by the machines.
4. *Selection and training.* Identify which workers have the skills or abilities to operate, maintain, and program the robots and which will need training to perform these tasks, then decide the kind and extent of training that will be needed.
5. *Maintainability.* Since most of the cost of maintenance is incurred in diagnostics, the persons performing the maintenance on the robot system should be those persons already familiar with the robot's environment who have received further training in robotics maintenance.

6. *Safety*. Protect the robots and the humans from each other by designing safe production layouts, educating workers about potential hazards, providing workers with protective equipment if necessary, developing clear safety guidelines, and assigning responsibility to a given department or group of workers for ensuring that safety regulations are followed.
7. *Management*. The decision to automate should not be made without the full knowledge, consideration, and cooperation of affected employees. Workers' resistance to changes in production and task assignment is lowered when such plans include their input. Some of the issues that should be considered in such plans include labor grades, pay scales, training, departmental transfer, and operator responsibilities.
8. *Communications*. Continuous, open communication between workers and management will ensure optimum implementation of the contemplated robotics system. Both formal and informal lines of communication should be established between those who are indirectly and directly involved in the automated process.

Howard emphasizes that these factors should be considered from the very beginning of considering a robotics system installation to the actual implementation of the system. They will continue to be important throughout the operation.

Another way to help ensure a successful installation is to follow Susnjara's (1982) five cardinal rules of robot installation.

Rule 1. Think simple. In considering the use of industrial robots in an existing plant, a survey of possible applications should be made. For the first robot installation, this first rule would advise us to go for a simple, proven application. For example, if the choice is between spot welding and assembly, the more proven spot-welding application should be selected. Thinking simple might also mean that the installation that has the greatest possibility for increasing profits might be delayed until the installation with the greatest degree of success has been made. Susnjara states that "a successful robot installation even in a simple application will quiet the skeptics, convert the doubters, quell most of the fears of labor and management people, and set the stage for future successes."

Rule 2. Be thorough. The purpose of this rule is to avoid "surprises" in the selected application. Perhaps a production technique with which everyone is familiar but that has not been considered in the initial survey will be encountered, or perhaps some unreported variation in the process to which the human operator has adapted in the past will be uncovered. Such surprises could determine the success or failure of the installation. To avoid these potential problems, Susnjara advises involving both the production employees who are intimately familiar with the process and the supervisors in the planning of the robot installation. Being thorough involves long-range planning and profitable use of the industrial robot. This includes many factors past the initial design and installation. Maintenance, training, and continued concern about the operation are essential.

Rule 3. Be reasonable. This rule involves an awareness of what the robot can and cannot do. Researchers are available to attempt the challenging and unsolved problems. Installation of a robot system for a given application should generally not be made until its efficiency has been proven. This rule also involves having reasonable expectations of the robot manufacturer. They cannot be expected to do detailed applications engineering work as part of the purchase price of the robot. Finally, being reasonable involves being realistic in expectations and planning.

Rule 4. Be honest. This rule applies to every phase of the robot installation, from the initial survey to analysis after years of successful use. It refers not only to technical appraisal but also to open and honest discussions with the people at all levels involved with the robot, and even includes community relations. The help and encouragement of management, employees, union representatives, and the community depend upon fair and accurate reporting of the robot's implications.

Rule 5. Be careful. Safety is always of prime concern and requires proper awareness by everyone involved. Training and retraining are also part of being careful. Casual familiarity with robot systems and procedures can lead to carelessness, with unfortunate consequences. Being careful with the computer programs by keeping current backup copies can also save weeks or months of effort from a system or medium malfunction that could destroy the information saved.

These five rules and perhaps others you devise help ensure the success of a robot installation. A machine as powerful and versatile as an industrial robot cannot be installed and forgotten. Constant training, retraining, and attention to safety considerations are necessary.

7.2 Economic Justification of a Robot

There are several reasons for installing an industrial robot. The increased productivity that can result from the more consistent operation and constant throughput of the robot is an important issue. Recall the popular folk song about the contest between John Henry, the steel-driving man, and the automatic spike-driving machine—no matter how hard or fast he worked, the machine outperformed him. A robot may be justified on productivity increases alone, since it is not limited to an 8-hour day. An increase in productivity can be measured in the number of parts produced per day. This factor alone may make it clear that a robot installation is justified.

Improved quality is another reason for installing an industrial robot. The consistency of the robot operation can be translated into consistency in the quality of the product. In conjunction with improved quality is the reduction of defects and scrap materials produced by the operation. For example, in a spray-painting operation, greater consistency may be realized in the form of more uniform thickness of the applied material. Since the layers of paint may be only one-thousandth of an inch thick, the application of a coating two-thousandth of an inch thick would be barely perceptible to a human operator but would use twice as much material. One manufacturer of food products installed a device for measuring the amount of cheese in a pound package of cheese. The savings to the manufacturer in not giving away more cheese than required and in reduced time satisfying the government inspectors paid for the machine in just a few months. Another company involved in manufacturing cathode-ray tubes was almost put out of business because of the uncontrolled method it was using to deposit phosphor on the faceplate. As the price of phosphor increased over the years, competitive companies automated the process and were able to offer a higher quality product at half the cost. Even in applications in which the robot is not considerably faster than humans, the consistency of operations may provide a definite advantage. For example, in die casting or plastic molding, the constant cycle time of the robot can permit the temperature of the product to stabilize and thus produce more consistent goods. Furthermore, the continuous operation of an industrial robot can eliminate the need for intermittent rest breaks that can result in inconsistent quality in a product. Welding is another application in which consistency can lead to considerable savings and prevent costly failures.

Improving the quality of work by eliminating health risks to humans in hazardous tasks is another major consideration in the use of industrial robots. For example, since the beginning of serious work with radioactive materials began in the 1940s, some type of remote manipulator or robotic device has been needed. Perhaps not so obvious but just as important are the many toxic environments encountered in spray-painting, chemical-processing, or other dusty manufacturing environments. One company we visited makes products from asbestos. Breathing microscopic asbestos particles can lead to severe lung disease. This company had spent over a million dollars installing ventilation equipment in a large high-bay factory to satisfy Occupational Safety and Health Administration (OSHA) standards. Looking back is always easier; however, it seems possible that a better solution might have been a confined space and robot installations. Many robots could be purchased for a million dollars.

Applications for robots to perform undesirable tasks due to difficult working conditions, such as noisy, dirty, or hot environments, those with noxious fumes, or simply fatiguing jobs requiring lifting heavy loads or working at a very fast or monotonous pace, might be more difficult to economically justify but have a significant humane value. Many manufacturing jobs fall into these categories.

Since the jobs are now being done by humans, there is an economic baseline for comparison with a proposed robot installation. In such cases, a careful economic analysis is required.

To illustrate how the many factors interact that determine whether a robot installation is a good investment, let's consider the following hypothetical example developed by University of Cincinnati Professor Ronald Tarvin. This economic analysis would logically follow a survey of a plant in which it has been determined that at least one potential robot application exists. For the economic analysis, we will be concerned with such factors as the payback period, the return on investment, a cash flow analysis, tax credits, depreciation, and the tax bracket the company is in.

Example Economic Justification. The installation of a robot would first require a capital investment. For this example, we will assume that sufficient capital is available to cover the initial costs. If the capital is not available, then the cost of borrowing the funds must also be included. Susnjara provides tables of the cost of capital borrowed at various interest rates for various periods of time that may be used to include this element.

Suppose that the items to be purchased include the following.

One industrial robot with options	$60,000
Gripper or process tooling	1,500
Safety equipment	2,000
Sensors and interface	1,250
Conveyors	1,750
Total capital investment	$66,500

These costs are typical of a robot installation. The price of the robot alone may only be $50,000. However, most manufacturers offer certain options useful in particular applications. For example, the software and hardware for tracking a moving conveyor may be required. The cost of the gripper or process tooling is also listed separately, since these come in such a variety of types. The safety equipment may consist of simply a chain barrier or perhaps a fence or wall. This is, again, very dependent on the application. Conveyors that may be needed to transport parts to and from the robot have also been included. This example illustrates that the capital investment required is more than the cost of the robot alone.

There would also be some expenses necessary for starting up the installation. We will assume the following.

Feasibility study (400 hr @ $20/hr)	$ 8,000
Engineering time (200 hr @ $20/hr)	4,000
Site preparation (80 hr @ $15/hr)	1,200
Installation (60 hr @ $15/hr)	900
Total installation costs	$14,100

The feasibility study would be done before installation but could be directly associated with it. The cost estimate is based on 10 weeks of engineering time to survey the facility and determine the most appropriate installation, talk to the employees, management, and union representatives, and provide a report. The engineering time includes 5 weeks to design specialized tooling, perform plant layout changes, fabricate accessories, and attend training classes on the operation of the robot. The site preparation would include the cost of constructing safety barriers, providing power sources, and other necessary renovations. Finally, the cost of installation, which would normally be done by the robot manufacturer personnel, is included.

Thus far, the robot has been purchased and installed at a cost of $80,600. Let's now consider how these costs might be offset.

First, let's assume a modest increase in productivity, perhaps due to the decreased break times no longer needed for the operation. Suppose eight more parts per day can be produced and that each part is worth $20. The annual increase is

$20/part × 8 parts/day × 250 days/yr = $40,000
 Total productivity: $40,000

This increase would double if a two-shift operation were used. However, for this example, we will consider only one shift.

There could also be a savings due to the robot shifting a worker to another operation in the plant. Suppose that the robot has taken over a hot, dirty, or difficult part of the operation and reduced the number of workers from five to four. The shifted worker would result in a cost savings for the operation of

$11/hr × 2000 hr/yr × (1.5 benefit factor) = $33,000
 Total labor savings: $33,000

The labor cost of $11 per hour for 2000 hours/year is multiplied by a fringe benefit factor to provide this total cost.

Thus far, the savings resulting from increased productivity and labor costs is $73,000. However, we will encounter some costs in operating the robot. If the robot requires 30 kilowatts of electric power purchased at 4 cents per kilowatt hour, the annual power cost would be

30 kW-hr × $0.04/kW-hr × 2000 hr/yr = $2400
 Total power cost: $2400

Periodic maintenance would also be required. We will assume the cost of this maintenance to be 3.75 percent of the equipment costs of $66,500, or

Total maintenance costs: $2500

The total operation expenses per year would be $4900.

Adequate portrayal of other factors, such as depreciation and tax deductions, requires a cash flow analysis. This is simply a tabulation of the costs encountered during the lifetime of the equipment. For this example, a 5-year time period is assumed for the depreciation of the equipment. Of course, the robot would still be functioning after 5 years. One expert estimates that the average lifetime of an industrial robot is 8 or 9 years. Many of the early industrial robots are still functioning after more than 10 years. There would also be a salvage value to the robot after 5 years, since it might contain some 6000 pounds of metal. However, for this example, the salvage value will not be considered.

We will also assume that a simple straight-line depreciation is used. That is, each year, one-fifth of the equipment cost of $66,500, or $13,300, may be used as a tax depreciation. We will also assume that the corporation is in the 50 percent tax bracket. This means that for each $2 profit, $1 will be paid in taxes, and that for each $2 deduction, $1 in taxes will be saved. Finally, an inflation rate of 10 percent/year is assumed. This affects the labor savings, which will be increased by this amount each year.

The total cash flow analysis is shown in Table 7–1. In the starting year, year 0, the initial investment for the equipment and start-up costs is made. In the first year, called year 1, the effects of this investment begin to appear. The first item listed is a 10 percent tax credit for investing in capital equipment. Also, the start-up expense has been listed as a tax deduction, which results in a tax savings of $7050, since the corporation is in the 50 percent tax bracket. The equipment depreciation is also used as a deduction, resulting in a tax savings of $6650. The wages and benefits of the displaced worker provides a savings of $33,000, but half this amount, or $16,500, must be paid in taxes. Similarly, the production increase of $40,000 is considered income, so half of it, or $20,000, must also be paid in taxes. The maintenance expense of $2500 is listed as a deduction, resulting in a tax savings of $1250. Similarly, the energy expense of $2400 results in a tax savings of $1200. The total cash savings for the first year is $54,400. In the following years, the main difference is that the labor savings is increased by the assumed inflation rate. Note that, although the initial cost of the robot and start-up expenses were $80,600, the total savings over the 5-year period is $254,828.

Financial analysis also looks at two other quantities that may be determined from

Table 7–1 Cash Flow Analysis for an Industrial Robot Installation

	Year 0	Year 1	Year 2	Year 3	Year 4	Year 5
Initial investment	−66500					
Tax credit (10%)		6650				
Start-up expense	−14100					
Tax credit		7050				
Equipment depreciation		13300	13300	13300	13300	13300
Tax cost		−6650	−6650	−6650	−6650	−6650
Wages and benefits of replaced worker		33000	36300	39930	43923	48315
Tax cost		−16500	−18150	−19965	−21962	−24158
Productivity increase		40000	44000	48400	53240	58564
Tax cost		−20000	−22000	−24200	−26620	−29282
Maintenance expense		−2500	−2750	−3025	−3328	−3660
Tax credit		1250	1375	1513	1664	1830
Energy expense		−2400	−2640	−2904	−3194	−3514
Tax credit		1200	1320	1452	1597	1757
Annual cash savings		54400	44105	47851	51970	56502
		Average annual savings = $50,966				

Source: Ronald Tarvin.

the cash flow analysis. One is called the internal rate of return on the investment, which for this example is 56 percent. Another is the payback period, which is the reciprocal of the internal rate of return and is 1.79 years for this example. This value is determined by equating the total initial cost to the sum over the number of years of the equipment's lifetime of the cash savings for each year, divided by 1 plus the internal rate of return on the investment raised to the power corresponding to the year index. This involves an iterative calculation best done on a computer. The internal rate of return on investment may be interpreted as the average percentage return on the capital investment. Most companies would consider a rate of 56 percent a healthy investment.

Several other factors, such as the savings due to less reworking of the product and reduced materials costs from less scrap produced, could also be included in the analysis. Also, other intangible benefits, such as reduced costs of OSHA compliance, improved utilization of floor space, increased machine cycle rates, quicker new-run changeovers, and greater system flexibility, could become important.

This example illustrates the economic justification of a single industrial robot. Let's now consider the economics of a work cell that includes a robot and a sensor system, such as vision sensors, to see how the costs and justifications might change.

Let's imagine that we had an application requiring a two-shift operation of a materials-handling robot and a vision system. Furthermore, in this application, the two workers currently doing this job were retained. One was retrained to operate and program the robot. The other was transferred to another location in the plant to keep pace with the increased production rate made possible by the robot.

To evaluate the economics of the robot installation, the following assumptions were made: a 5-year, straight-line depreciation; a 5 percent rate of inflation per year; a 50 percent corporate tax rate; and no salvage value of the equipment after 5 years.

The initial investment expenses include

Industrial robot	$66,800
Gripper	4,000
Safety equipment	4,000
Vision sensor	40,000
Conveyors	5,000
Other fixtures	5,000
Total capital investment	$124,800

The start-up expenses are

Feasibility study (400 hr @ $20/hr)	$8,000
Engineering time (200 hr @ $20/hr)	4,000
Site preparation (80 hr @ $15/hr)	1,200
Installation (80 hr @ $80/hr)	1,600
Total installation costs	$14,800

In this application, no savings due to wages and benefits of replaced workers results. However, a considerable savings resulted from an increased production rate. In particular,

Parts per day increase	20
Working days per year	200
Amount per part	$50
Total productivity savings: $200,000	

The energy expense of the new equipment is

30 kW-hr $\times$ 0.04/kW-hr $\times$ 4000 hr/yr = $4800
   Total power cost: $4800

For periodic maintenance, we again assume the cost to be 3.75 percent of the original equipment cost, or

$124,800 $\times$ 0.0375 = $4680
   Total maintenance cost: $4680

In this example, we also include an equipment insurance expense of 5 percent of the equipment cost, or

$124,800 $\times$ 0.05 = $6240
   Total insurance cost: $6240

We may now determine the cash flow analysis for the industrial robot installation. This is shown in Table 7–2.

The internal rate of return on the investment is again calculated by equating the initial investment of $138,800 to the sum of the cash savings per year, divided by 1 plus the internal rate of return raised to the power of the year.

The following BASIC program was used to calculate the internal rate of return, which was 80 percent, and the payback period of 1.24 years.

```
10    REM INTERNAL RATE OF RETURN
15    EMIN=999999
20    FOR I = 0 TO 1 STEP 0.001
30        E=139600-124500/(1+I)-109227/(1+I)**2
40        E=E-114064/(1+I)**3-119143/(1+I)**4-124475/(1+I)**5
50        IF ABS(E) < ABS(EMIN) THEN IR=I
60        IF ABS(E) < ABS(EMIN) THEN EMIN=E
70    NEXT I
80    PRINT "INT RATE OF RET";IR; "PAYBACK PERIOD";1/IR
90    END
```

Table 7-2 Cash Flow Analysis for Robot Visual Inspection Example

	Year 0	Year 1	Year 2	Year 3	Year 4	Year 5
Initial investment	−124,800					
Tax credit (10%)		12,480				
Start-up expense	−14,800					
Tax savings		7,400				
Depreciation		24,960	24,960	24,960	24,960	24,960
Tax cost		−12,480	−12,480	−12,480	−12,480	−12,480
Wage savings		0	0	0	0	0
Productivity increase		200,000	210,000	220,500	231,525	243,101
Tax cost		−100,000	−105,000	−110,250	−115,763	−121,551
Maintenance expense		−4,680	−4,914	−5,160	−5,418	−5,689
Tax savings		2,340	2,457	2,580	2,709	2,844
Insurance expense		−6,240	−6,552	−6,880	−7,224	−7,585
Tax cost		3,120	3,276	3,440	3,612	3,792
Energy expense		−4,800	−5,040	−5,292	−5,557	−5,834
Tax savings		2,400	2,520	2,646	2,779	2,917
Annual savings	−139,600	124,500	109,227	114,064	119,143	124,475

Average annual savings = $118,282

7.3 Economic Justification of a Work Cell

The industrial robot often serves as the central element in an automated work cell in which an entire product or assembly is produced. We will now briefly describe an example justification developed by Holmes (1979) to illustrate the economic justification of an entire work cell.

A comparison will be made between a manufacturing cell consisting of two manned CINTURN turning centers (TC) and one in which the identical turning centers are loaded and unloaded with a Cincinnati Milacron T3 robot and a single operator. The comparison will be made on a two-shift operation basis for a 1-year period. A table of the productivity factors for the two alternatives is given as follows.

	Human-operated	Robot
Available cut time	120 min/hr	120 min/hr
System attention	9 min/hr	12 min/hr
Efficiency	80%	90%
Total utilization	88.8 min/hr	97.2 min/hr

The available cutting time is simply the sum of the cutting times for both machines.

Loading and unloading the systems are reflected in the system attention times. The overall efficiency of the robot process is slightly greater than for the human-operated centers because of the greater consistency of the robot. The total utilization factors are determined by subtracting the attention time from the available time and then multiplying by the efficiency.

The next set of factors relate to the times involved in actually making a part.

Part cycle time (min)		
Load/unload	1.29	0.37
Cutting time	2.50	2.50
Total time (min)	3.79	2.87

Note that the robot can load and unload the turning center slightly faster than a human because it is in a fixed location. The cutting times are identical for both.

Next we include a fatigue factor.

Fatigue factor	1.04	1.00

This factor simply reflects the consistency of the robot compared with that of a human. The total time for producing the product can now be determined by multiplying the total machine time by the fatigue factor.

Total part time	3.94 min/piece	2.87 min/piece

We may now determine the throughput, or the number of pieces that can be produced per hour.

Throughput: Human-operated 22.5 pieces/hr; robot 33.9 pieces/hr

We now have enough data to determine the increased productivity, which is simply $(33.9 - 22.5)/22.5$, or 50.7 percent. This increase in productivity is as much as could be produced by an added turning center.

Let us now consider the factors that will lead us to the return on investment of the robot center. We will assume a labor cost of $8.25 per hour. Other factors, such as taxes and insurance, maintenance, energy, and tooling costs, will also be tabulated.

Annual operating cost, attended	Three-TC operator attended	Two-TC robot
Labor cost/year	$46,530	$15,510
Taxes and insurance	13,465	14,100
Maintenance	21,500	22,500
Power cost	10,500	8,800
Annual fixed tooling cost	1,500	1,000
Total	$93,555	$61,910

The equipment costs for three manually operated versus two robot turning centers will now be totaled.

Equipment	$573,000	$600,000
Installation (80 hr @ $7.5/hr)	600	600
Total	$573,600	$600,600

The tax factors will now be added.

Investment tax credit (10% of capital equipment)	$57,300	$60,060

Maintenance will again be assumed to be 3.75 percent of the equipment costs.

Maintenance	$21,500	$22,500

Energy costs must also be added.

Energy costs	$10,560	$8800

Finally, taxes and insurance will be considered to be 2.35 percent of the capital investment.

Taxes and insurance	$13,465	$14,100

Enough information is now available to determine the payback period and rate of return on investment. The total equipment costs are $516,012 for the three turning centers versus $540,312 for the two centers and robot, for an increase of $24,300 for the robot system. However, the annual operating costs are $93,555 for the human-operated center versus $61,910 for the robot center, for a savings of $16,455 for the robot system. The cash payback period is 1.5 years, and the rate of return on investment is 65 percent for the automated work cell.

This example illustrates that the costs involved in a robot-centered work cell are substantial; however, the rate of return on this capital investment is quite attractive. Many costs not directly associated with the robot must be considered in the economic analysis. A combination of savings due to productivity increases, labor savings, increased efficiency, and greater cutting time make the robot work cell an attractive method.

An operator is still available for the robot work cell. The human's adaptability is now used to greater advantage in such tasks as ensuring that the raw materials supply for the cell is available, monitoring the process for tool breakage or other problems, and coordinating the off-line support to ensure the successful operation of the system.

7.4 Economic Considerations for the Automated Factory

We have now considered the justification of a single robot and an automated work cell. Recall that the capital investment for the single robot was about $80,000, but that for the automated work cell was about $540,000. We might expect that the costs of an automated factory will be substantially larger.

The factory of the future will include all the functions that must be accomplished in a discrete product-manufacturing plant, such as parts transport, inventory storage, processing, assembly, and inspection. However, these automated functions will be integrated into a consolidated system. A distributed computer information and control system will automatically route the raw materials through the plant to the warehouse. Robots will be used in materials handling, painting, welding, machine loading and unloading, and all major manipulative actions. Customer order information can be used to determine the proper routing of materials to fill the order. Inspection work will be done by industrial robot systems equipped with vision, tactile, and other sensors. Inspection will be done on a 100 percent basis rather than by sampling a small percentage from each batch. Humans will still be required to perform the management, operation, and maintenance functions.

Shunk et al. (1982) suggest that the factory of the future must be defined in terms of its systems. That is, the factories will be groups of manufacturing cells that work in conjunction with each other to perform the entire production's operations. Whereas most industries today use "islands of automation" in conjunction with human or fixed automation operations, these new factories will be composed entirely of flexible automation cells. The key to the success of these factories will be their tremendous potential for flexibility, efficiency, and effectiveness. The concept of such a factory must involve both hard-technology and soft-technology considerations. The hard-technology view concentrates on the actual production of the product; the soft-technology view concentrates on the communications needed for monitoring, controlling, and reporting the condition of the systems. In making such systems mesh in a successful manner, a critical factor is the amount of user involvement in defining the systems. Designers and potential users must be able to communicate effectively about such things as what functions are to be performed by what (or by whom), how the systems will interact, what information will be needed, and how people will need to interact with each system. In determining such systems, a simulation is strongly recommended. Computer-aided simulation can allow both designers and users to see how the systems will work before they are built, eliminating the need to redesign or scrap the systems after they have been built.

Questions

1. Determine the internal rate of return on the investment and payback period for an industrial robot system that requires an initial investment of $137,600 and produces annual cash savings in the first 5 years of $119,650, $109,833, $114,644, $119,696, and $125,001, respectively.

186 Economic Considerations and Justification of Industrial Robots

2. Determine the internal rate of return on the investment and payback period for a robot installation that requires an initial equipment investment of $65,200 and start-up costs of $14,400 and produces annual cash savings in the first 5 years of $54,400, $44,105, $47,871, $51,900, and $56,502, respectively.

3. List five benefits that frequently are achieved through the application of industrial robots.

4. If an industrial robot costs $53,000 and a 5-year straight-line depreciation with a first-year 20 percent acceleration, what is the total first-year depreciation?

5. In the first installation of an industrial robot at your plant, there seem to be several alternative courses of action. Discuss the prudence of the following options: building your own robot; selecting a proven task and application; selecting the task with the highest production volume; selecting the lowest priced available system; and disregarding the human relations aspects.

6. The initial investment and start-up expenses for an industrial robot installation do not include the cost of retraining. Why? Where should this cost be covered?

Social Impact of Industrial Robots

As we have seen in Chapter 7, robots are becoming more economical as labor costs rise, and the price of robots is likely to decrease over time as well. Because of this, according to most predictions, there could be 1 million robots in use in the United States alone by the end of the century (Gore, 1982). Therefore, robots should be of interest and concern to workers whose jobs are likely to be overtaken by robots and who must be retrained to do other jobs, managers and supervisors whose duties will probably undergo great changes, educators who must prepare tomorrow's work force, and students, who are the work force of the future. In this chapter, we will examine some of the effects that automation has had on society in the past and is likely to have in the future, some of the social problems present today in the work force, and how the widespread implementation of robots in society is likely to affect those problems.

One of the main distinctions between a robot and any other piece of automated equipment, as we have specified, is that its function or task may be easily changed or modified, with only software modifications instead of special, extensive retooling. Because robots are usually more complicated than other automated machinery, involving more complex software and hardware knowledge, technicians who are specially trained to work with robots will be in high demand, once large numbers of robots are in use. One estimate projects that as many as 800,000 jobs will open up for robot technicians during the next decade (Cetron, 1983). However, there is general disagreement in the robotics community over where these robot technicians will come from. Some argue that many of these technicians are likely to be graduates of 2-year training programs or re-educated skilled workers. Others insist that the majority of these jobs will be available only to engineers with special training in flexible automation systems. There is already a large demand for such engineers. Still other roboticists predict that robots won't need to be repaired—they may simply be replaced with newer, better robots when they break down. However, this does not eliminate the need for general maintenance of robots. As the old saying goes, everything put together sooner or later falls apart. The only question is, who will put everything back together? So much disagreement exists over this question that it is risky to make hard predictions at this time. It does seem likely, however, that there will be jobs for the 2-year technicians if their training is sufficiently both specific and broadly based.

Robotics will create entire new industries, as well as put some industries out of business, which will change the type of jobs that will be widely available. These new industries will be those that make products that relate to robot use, such as new types of conveyors, end tooling, imaging systems, positioning devices, and transport mechanisms, as well as those that make robots and/or robot components. Consequently, the kinds of jobs that robots and computers will be opening up in the next 20 years include those for software experts, simulation designers, computer programmers, robot retrainers, artificial intelligence engineers and scientists, automatic factory security experts, educational and career counselors, human/machine interface experts, and employee relations counselors and trainers, as well as a wealth of jobs in robotics sales, marketing, installation, control, maintenance, repair, and education. The kinds of jobs that are likely to gradually disappear over the same time period are those for machinists, machine loaders, finishers, welders, packers, citrus pickers, and warehouse workers. Those who are at present involved in factory and manufacturing manual labor will be those most affected, but some "white collar" jobs, such as typists, librarians, clerks, operators, and graphic artists, are also likely to taper off before 1990.

As more robots become used, the need to understand and utilize robots will extend to disciplines not traditionally technical, such as economics, physiology, psychology, and sociology (Sitkins, 1983). Still, one of the most significant impacts of robotics and modern automation will be on education and training.

8.1 New Requirements in Education and Training

Many of us mistakenly assume that robots will permanently eliminate the need for people in the workplace. Nothing could be further from the truth. Robots are machines and, like all other machines, will need people to design, build, program, install, troubleshoot, supervise, maintain, and repair them. Each job will require a different degree of skill and type of knowledge. At present, the robot-manufacturing community needs many qualified robotics applications engineers. A robot is a wonderful tool in the hands of such trained engineers, because the robots are available and the potential applications almost unlimited—these engineers simply bring the two together. Also, most robot users demand the more highly trained people to design, implement, and engineer their installations. However, after large numbers of robots have been applied, there will be many opportunities for nonengineers to become involved with robots. In fact, some roboticists believe that the majority of jobs that will be created by the widespread use of robots will be offered to technicians with training approximately equivalent to, say, a present-day machinist or computer programmer. These technologists-technicians will serve as liaisons between engineers and applications and also be required for regular servicing of installed robot units. It is probable that robot technicians will assume many tasks that have formerly been performed by engineers (Fitzgerald, 1983). Thus, most robotics experts repeatedly emphasize the need for training programs to adequately prepare the labor force for these new jobs.

It is difficult, however, to establish just what kind of training is needed, as well as how many technicians will be needed. For this reason, a good deal of industrial and educational cooperation will be necessary to minimize the mismatching of training and available jobs. For example, there is a current shortage of applications engineers, because, apparently, no one foresaw the number of such engineers that would be needed. The robotics and related technology industries must work closely with educational institutions to try to prevent such shortages, and to prevent oversupply of technicians, for example. Students also have a responsibility to watch the job market and to adjust their curricula accordingly, with the help of their advisors and faculty, so that their skills and knowledge will be in demand when they graduate. These educators and advisors must, therefore, be aware of the impact robots are making and will make on the job market.

As robot manufacturers, educators, and users come together in workshops and seminars, the understanding of just exactly what kind and how many technicians are needed will develop, enabling both the schools and their graduates to perform a needed service without false expectations or needless course work. For example, Weisel (1983) lists seven categories of job assignments in the study, manufacturing, and applications of robotics—research and development, design, manufacturing, application, marketing, service-maintenance, and training-teaching. The potential employers are robot users, robot manufacturers, educational institutions, and robotics-related industries.

Most colleges and universities offer robotics programs and courses within existing engineering departments; 2-year educational institutes generally offer complete, 2-year programs in robotics. All engineering disciplines are interested in robots and their applications, but many other disciplines are also beginning to be interested. Difficulties can arise in developing robotics courses that cross departmental and college boundaries. There is also a recognized need for robotics and manufacturing courses for those in business colleges who plan to become managers and planners. Indeed, robot awareness courses could be beneficial to students in every discipline.

Robotics International of the Society of Manufacturing Engineers (SME) reported in its *1983–1984 Directory of North American Robotics Education and Training Institutions* (Berg, 1983) that there were twelve 2-year schools (six located in Michigan) that offered either robotics degrees or options in robotics as part of an engineering degree. There were fifteen 4-year schools that did so. Robotics courses were offered in twenty 2-year schools and thirty-five 4-year schools. In only 1 year, however, these figures more than doubled. The *1984–1985 SME Directory of Robotics Education and Training Institutions* lists over 340 campuses across the country that now offer robotics courses: 7 universities now offer doctorates, 44 offer master's degrees, 70 offer bachelor's degrees, and 181 offer associate degrees. This is an indication of the widespread interest being generated in the study of robotics.

Major robot manufacturers, such as Cincinnati Milacron and General Electric, have been instrumental in helping set up educational facilities at 4-year institutions by donating robots to colleges and universities or by making educational packages available at a low cost. Such programs are extremely valuable in providing students with an opportunity to have hands-on experience to supplement their theoretical training. Experience with actual robots is necessary for potential applications engineers. Where robots are not available for such programs, some robot manufacturers and schools have developed cooperative efforts, such as work-study programs.

In addition to assisting educational programs, robot manufacturers themselves also offer introductory robotics courses, especially in the use and care of their line of robot systems. However, these programs can be quite expensive, costing up to $1000 per week per student. Also, factories are not generally considered ideal classrooms, nor are the instructors specifically trained to teach. Still, these programs fill the gap left by inadequate training elsewhere, and most do provide excellent training. An example of such a training center at a factory is shown in Figure 8–1.

Many major technical societies are taking steps to help their members become aware of the robotics evolution through symposia, seminars, workshops, and conferences. Such organizations as the American Society for Engineering Education, the Institute of Electrical and Electronic Engineers, the Institute of Industrial Engineers, the

Figure 8–1. A robot training center. Note that several robots are available for hands-on operation by students. Also, various practical examples, such as spot and seam welding or conveyor tracking, are set up for student use. (Courtesy of Cincinnati Milacron.)

American Society of Mechanical Engineers, and the International Society of Optical Engineering are placing special emphasis on robotics. The Society of Manufacturing Engineers has an individual member association called Robotics International that is dedicated to the dissemination of technical information to robotics students and professionals.

8.2 Effects of Robots on Employment

The first questions people usually ask about robots are, "Will robots take over my job? Are they going to cause unemployment?" Since many factors contribute to or aggravate unemployment, these are not easy questions to answer. However, we may examine examples of past mechanization to determine the effects of the widespread use of robots in the work force. We may also consider humans and robots as valuable resources, neither of which should be underutilized.

Unemployment is defined in Webster's as the "involuntary idleness of workers." The Office of Technology Assessment, in a recent Congressional report (1982), makes a distinction between short-term unemployment and persistent job loss, which will be described later. It also distinguishes between job loss and displacement or job shift. For some time, it has been argued that more jobs are created than are eliminated by implementing high technology. However, if the jobs created require new skills, retraining programs will be necessary to prevent permanent displacement of workers in those industries that use robots. In fact, it has been argued that the single most important impact of robotics on society will be this retraining of many people to do new, unfamiliar work.

This retraining, at first consideration, may not appear to be a problem. However, for various reasons, some workers may be reluctant to retrain for a new job and may resent the machines that make their old jobs obsolete. However, there is tremendous pressure on certain industries, such as the automotive, to modernize their processes by using the latest technology available to remain competitive in the world market. Often, these industries face little choice in the matter of robotization. In the past several years, many markets in the United States have been lost because U.S. companies could not remain competitive with goods produced by such countries as Japan, a heavy user of robots (some 32,000 by the RIA definition to our 6300 in 1982), or by countries whose laborers are paid very low wages. There is no fixed quantity of work available to people. Advances are needed to improve the quality of life in every field of endeavor, from the cures for cancer in medicine to the ideal home. There is, however, a fixed quantity of resources available to perform this unlimited work. Knowing and using these resources wisely are essential to our continued well-being.

One of our most precious resources is human creativity. We should not have to waste it on tasks that can be performed by machines. This is not to say that small, individually owned businesses and occupations will not remain an important element of our society. Craft and artistry will continue to be important expressions of human thought, skill, and creativity, and their products will always be in demand. However, in the markets in which machine quality is adequate and the mass production of items is essential to supply the demand, we must use both human and machine resources as efficiently as possible to be competitive with other countries.

The *Robot Times* (1983) reported that Donald Smith, director of industrial development at the University of Michigan, projects that, since one robot has the potential of displacing two workers, extrapolating this into 100,000 robots by 1990 could mean the displacement of 200,000 workers. He estimated that 90 percent of these displaced workers would be transferred to other jobs within the user industry, that 5 percent would retire, and that only 5 percent were likely to face permanent termination of employment. However, he suggested that this 5 percent could be absorbed by the new jobs that will be created by the growth and expansion of the robot industry. For example, as we have already mentioned, a considerable number of robot applications engineers would be needed, along with sufficient robot technicians and operators to maintain, operate, program, supervise, and repair the robots in use. Other jobs might well result from more U.S. dollars being spent in the United States and will vary over every opportunity for employment. The severe unemployment of the past decade was not caused by industrial robots. Rather, lack of automation contributed to the slump in productivity, growth, and employment. Furthermore, unemployment improved in 1982, a year in which there were approximately 6300 industrial robots in use in the United States, but unemployment remained at intolerable levels in earlier years, years in which far fewer robots were used.

The prediction that we will have 100,000 industrial robots in use by 1990 assumes full production by every robot manufacturer in the United States, as well as purchasing robot imports, just to achieve this number. Therefore, this estimate may be somewhat high. However, several experts project the year 2000 as one in which 1 million industrial robots could be in use worldwide. By this time, robots will not only be used in industry, but also in medicine, by the military, and in the home. Domestic robots that do such work as mowing the lawn and cleaning the floor would cause little loss of jobs but have the potential to significantly increase the disposable time of their owners, freeing them to do other work that brings higher rewards.

Total employment is an ideal, or a figure of tolerance. For example, "full employment" used to mean that only 3 or 4 percent of the labor force was out of work. This small percentage was considered tolerable and possibly inevitable, due to workers' moving, taking leaves of absence, or being between jobs, for example. Because of changes in the economy, however, the definition of "full employment" may mean that 6 or 7 percent of the labor force is out of work (Gore, 1983, p. 1). In our working population of roughly 80 million, this translates into the possibility of some 6 million people out of work. However, perhaps half this percentage will be made up of workers whose old jobs have been eliminated by modern automation and who are undergoing retraining for new jobs. This would relieve structural unemployment, the mismatching of skills to available jobs. Some experts predict that technology is changing skill requirements so rapidly that most workers will eventually undergo periodic retraining and education just to retain competency in their present jobs.

Many suggestions of how to match training to available jobs have been proposed. One is to charge the government with the job of using the information it obtains through various sources to create a job bank (Wright, 1983, p. 5). These data could then be made available to educational and training institutions so that appropriate counseling and curricula design could take place that would match student skills to available work. Another remedy might be to require industries that replace workers with robots to give those workers free retraining at the schools of their choice. The most feasible suggestion, however, is to do what is already being done by most industries—training the displaced workers themselves to take new positions within the industry.

It should be repeated that total employment is an abstract, not a given quantity. It can always be achieved, whether by a productive or nonproductive society, by simply enforcing "make-work" programs or by giving industries incentives to "overemploy," that is, to use more employees than are needed for maximum productivity. The current rate of unemployment would seem to be related to productivity growth, since the productivity slowdown in the United States coincided with the highest average unemployment rate since the Great Depression. Economic growth depends to a great extent on productivity advances. Since we have introduced robots into our economy, the unemployment rate has fallen to 8.8 percent, the lowest in 10 years. This does not mean that robots caused the unemployment rate to fall, but that improved technology has resulted in growth in productivity, with the result that more employment is available.

Some changes in old jobs and the creation of the new jobs that will be offered in all manufacturing plants have been projected by the U.S. Bureau of Labor Statistics for the year 1990. It shows the following shifts and changes in employment expected to occur compared with 1978 figures (Guterl, 1983).

Machine tool operators	12% decrease
Machine setup workers	24% decrease
Tool and die makers	24% decrease
Maintenance electricians	28% increase
Computer service technicians	93% increase
Electrical engineers	27% increase
Industrial engineers	26% increase

These figures indicate that many jobs will be created from increased automation to offset the jobs that will be eliminated from society.

8.3 Effects of Robots on Workers

The people on whom robots are likely to have the greatest impact are the manual laborers in our factories and plants. There are certain things that robots can do much better than humans because they are more suited to the task. For instance, we do not need all the wonderful sensory, creative, and perceptual capabilities of a human to perform the job of spot welding. Most of the human workers who do this job become very bored with the task. A robot can perform this job more efficiently, more consistently, and more safely than can humans. If there were no jobs that robots could not do better than humans, then of course robots would not be needed. Since there are, it is prudent to look into the matter of what work is suitable for humans and what work is suitable for robots.

These determinations may be optimally performed by human factors experts. Human factors is a field of study that analyzes the specific ways humans and machines can work together, considering the unique capabilities and limits of each, and determining which offers the maximum utilization and efficiency in which jobs. Human factors engineers are often employed by manufacturers who plan to use robots in conjunction with humans on, say, an assembly line. Safety, for example, must demand prime consideration, for both the robots and the humans need protection from errors, mistakes, or sabotage. The speed at which the conveyor belt moves must be that which maximizes the capabilities of both human and machine without overtaxing either. Also, the work envelope and positioning of equipment or parts must be carefully analyzed, along with careful measurements of the robot's movements. This may not sound difficult, but it is very hard for a person to think like a robot. For example, it is simple to teach a human to pick bruised or withered apples off a conveyor belt. It is so easy that we fail to appreciate the many subtle processes involved in the ways humans perceive the bad fruit and reach out to remove it from the belt. If a robot is to be used for this job, it involves a basic intelligence and sensory system for recognizing and differentiating bad fruit from good. Then, each move the robot manipulator must make from starting position to finish must be carefully thought out. Robots cannot move the same way humans do. They don't automatically recognize obstacles in their way. They can't tell where the disposal bin is unless you tell them in programming terms. They can't just reach out and get the fruit, either. They must start from a hovering position, lower the gripper, sense when the gripper has sufficient traction on the fruit, lift the arm vertically, move to the location of the disposal bin, and so on. People working around the robot must know all its movements, its reach limit, where its manipulator will be during each step of the process, and other factors, to avoid accident. One suggestion is that all robots used in conjunction with humans in a process have a sensor capable of shutting down the system when an unexpected object, such as a human, invades the work envelope. Safety engineers work with human factors engineers in considering such possibilities and help in designing the work space so that the chance for accidents is minimized.

Psychologists and sociologists also sometimes work with human factors engineers to explore the attitudes of employees toward robots, the kinds of work the employees prefer, to determine which jobs are best performed by humans and which by machines in terms of satisfaction and utilization of talent or ability. Also, humans have a persistent view of robots as mechanical people, an attitude that must be taken into account when integrating robots in the workplace. Furthermore, employees should not be expected to learn about robots from experience; they should receive at least a basic introduction to robotics. Such knowledge will tend to eliminate much or all of the misinformation, fears, and misgivings they may have concerning robots.

Underemployment and Job Satisfaction

The word *underemployment* is used to refer to the situation that exists when a job fails to fully utilize a person's skills and/or abilities. Worker alientation refers to the worker's feeling that his or her job is meaningless, or that he or she has little or no control over the job. A special task force assigned to evaluate the problem found that there were high dissatisfaction and negative attitudes among workers who had to perform dull, repetitive, and seemingly meaningless tasks (Robertson, 1980). Yet, technology, bureaucracy, and the trend toward large organizations that break work down into many such tasks have made a large number of these jobs available to people. Although this increases productivity and control, it decreases worker satisfaction and utilization of human abilities. It also makes it impossible for some of the people employed in these jobs to derive satisfaction, other than pay, from their work, or to gain a sense of identity. When this happens, work becomes something that a person does only to earn a living, and their sense of identity must come from something they do after work hours. Such workers have little incentive, other than pay, to encourage them to stay with that job or to put any more effort into it than is required. This can lead to frequent job turnover, large numbers of scrap or rejects, and high absenteeism.

The human need for satisfying work is well established in the American community. In our country, particularly, work has always been a morally valued activity and the principal source of a person's identity and self-esteem. For example, the second question we usually ask of a new acquaintance is, "What do you do?" A person who inserts pins in shirts on a factory assembly line may have a hard time coming up with a job title that will give the new acquaintance any idea of the kind of person he or she is, or what skills and knowledge are possessed. This is generally true of assembly line workers. However, assembly line workers are not the only ones who are unhappy with their work.

A study by the University of Michigan Survey Research Center found that, for most people, the most important aspect of their work is how interesting it is (Robertson, 1980). These findings indicate that, of all workers, people over 55 and professionals get the most satisfaction from their jobs. White collar workers in general were twice as satisfied with their jobs as blue collar workers. The dissatisfaction of workers in general was most often attributed to low pay, inadequate supervision, and unpleasant working conditions. Other studies show that higher instances of poor mental health brought about by low self-esteem are found among industrial workers. Job satisfaction has also been shown to be one of the single strongest predictors of longevity. It is not, as you can see, an insignificant issue.

In our industrial society, three categories of work are available to people. These are primary industry, which refers to the work of extracting undeveloped natural resources; secondary industry, which involves all the work needed to turn these natural resources into usable products; and tertiary industry, which is better known as the service industry. When our country was young, most of our people were involved in primary industry work, such as farming, forestry, and mining. Even as late as 1900, the majority of people were employed in some sort of primary industry, principally agricultural. As our country was industrialized, more and more people were drawn into the secondary, or manufacturing, industry. These jobs were cleaner and more pleasant than those in the primary industries. A sort of mass exodus of people from rural communities to towns and cities occurred. Although rural people were traditionally self-sufficient, those in cities were more interdependent. Thus, the rapid expansion of urban communities created a massive need for many new services. By 1950, the service industry had outstripped both primary and secondary industries. As automation progressed, the number of jobs available in the manufacturing industry began to fall even faster. The Bureau of Labor Statistics reported that, between 1978 and 1983, the number of jobs in manufacturing (secondary industry) fell by nearly 2 million, but about 3.5 million jobs were created in service (tertiary industry). Primary industry employs only about 4 percent of all working people. This trend toward service-oriented jobs means that the prime qualifications for jobs is shifting from physical power to mental power. More jobs require a college education, and fewer require physical strength.

Automation and urbanization were responsible for this job shift from manufacturing to service, but this did not mean people had no work to do. It just changed the nature of their work. Many people feared that computers would put people out of work, but they didn't. They created many more jobs and new kinds of jobs, such as data processors, programmers, systems analysts, and technicians. The managerial, technical, and professional jobs that have evolved have had a tremendous impact on our occupational structure and the way we perceive work. Robots are likely to have a similar effect on our society. Many different kinds of jobs will be created from their use, and most of these jobs will be in tertiary industry. What does this have to do with job satisfaction and worker alienation? A special task force was assigned to study the problem by the Secretary of Health, Education, and Welfare. They reported that job satisfaction was highest for those in the tertiary segment of the work force. The reasons for this are many. For instance, these jobs are usually clean, offer pleasant working conditions, and are more versatile and interesting than those in the other industries. It is significant that the highest satisfaction was reported by urban university professors and the lowest was reported by unskilled autoworkers (Robertson, 1980).

The significance of understanding what kinds of work are available to people and how much satisfaction can be derived from them is that most of the robots used today are taking over those tasks that people disliked in the first place. They are not invading the tertiary industry to any great extent, and they are displacing humans mostly in those jobs humans find most dissatisfying or boring. It takes a skilled robot to do an unskilled human task, but even an unskilled human can do things no robot can do. That is why many economists predict a surge in minimum-skill jobs, such as those offered in fast-food restaurants, in the near future. However, as manufacturing begins to implement more robots, it may change the way employers perceive employees. They may see them as resources unsuited to monotonous, mechanical functions and more suited to those jobs requiring uniquely human abilities. As intelligent robots are developed, humans are likely to become even less valued for their physical attributes and more for their mental capabilities. Robots will undoubtedly contribute to the continuing shrinkage of jobs available in the manufacturing industry, but these jobs have been steadily declining for the past 40 years without robots.

Even if robots are not used, the number of these kinds of jobs will decrease. With our current population of job seekers, greater numbers of people will be seeking those shrinking jobs, so that fierce competition is likely to result. We have already seen some of these effects during the high unemployment of the early 1980s. One way to alleviate this problem is to upgrade the type of training and education our youth receive, so that large numbers of new young workers do not enter the job market unable to compete for the higher paying, more mentally demanding jobs. At present, large numbers of students are graduating who are qualified only for the jobs that robots or other machines are replacing at a rapid pace. Certainly, both national and local government should give some attention to the needs of society and help design educational and training curricula that ensure the proper training of its citizens. A high-school education is no longer generally considered adequate preparation for a secure future. Even a general college education is not always enough. The point is, our education and training programs will have to change and adapt to our changing, ever more technical world if we are to prevent mass unemployment.

In Japan, where industrial and manufacturing workers are regularly replaced by robots, there has been no mass unemployment. Why? There are many reasons for this; however, the one most often cited is that the labor force in Japan has received a basic education that emphasized science, mathematics, and technical literacy. Employers have a greater latitude in retraining these workers, and the retraining period is short. Furthermore, large organizations in Japan usually hire an employee for life, retraining him or her when new technology assumes the employee's task. Some reports have indicated that Japanese industries also at present tolerate a high degree of overemployment. In America, where high-school students can graduate without a single course in computer science, calculus, physics, or chemistry, the retraining is more extensive, expensive, and time-consuming. Industries are not ideally set up for educating or re-educating people, yet because many of our educational institutions do not have the needed resources, industry is stuck with the role. Also, most of our large organizations do not guarantee their employees a lifetime of employment. However, all the robot manufacturers we have talked to emphatically recommend to buyers that no employees be laid off as a result of a purchase of their robots, if it is at all possible. Most of these manufacturers offer extensive training courses to the buyer's employees in the use and care of their new robots as part of the sales package. This training is expensive, however, and is only a halfway measure to combat the problem of the basic lack of technical knowledge among industrial leaders and employees. Lamented one robot-manufacturing instructor, "I wish I could devote some time to taking our robots into the schools in this community. But I am kept busy at least 4 days a week just educating potential buyers in the most general information on robots. Everybody wants to know about robots, but I am only one person. This kind of basic education needs to begin in grade school."

Robot salespeople we talked to agreed that lack of technical knowledge on the part of potential users is one of the biggest factors holding up robot sales. An information specialist with Cincinnati Milacron defined three distinct groups of people in manufacturing and industry, each with different information needs. The first group is top and upper management, such as company presidents, chief executive officers, and upper level managers. They must be educated first, because even if a company's engineers appreciate and advocate the application of robots, this first group does not know enough about robotics to justify a robot purchase. They need the most general kind of information on robots, such as the kind of work they are and are not able to do. The most important aspect of robotics to this group of people is how robots can increase their profits, help them stay competitive, and become more productive, what kinds of problems they may create, and how to deal with those problems. In general, if utilizing robots cannot be shown to make the company a profit, then there is no point in educating the next two levels, except in considering long-term goals. The second group is composed of middle management, such as divisional managers. They need more specific information on robotics, particularly how robots can be applied to their particular industry or business. The third group consists of engineers, such as design and project engineers. This is the only group that really needs "hard" information on robot capabilities and features. These three groups form a hierarchy of information needs. Educating the second two groups in the uses and advantages of robot systems is largely futile if the first group, which makes all the final decisions, cannot be convinced that robots offer any advantages. For this reason, several manufacturers now offer "management" seminars aimed just at this decision-making group. This is expected to bring about more rapid dissemination of knowledge of robot basics in the near future.

It is important that all levels of decision makers in manufacturing receive a basic robotics education, because U.S. factory floors are responsible for producing about two-thirds of our country's wealth (Truxal, 1983). Corporate managers generally have business degrees with little knowledge of the technology that could significantly contribute to raised productivity. Engineers generally know little about manufacturing and do not always know how to apply their knowledge to such problems as numerical control. Those who do understand the manufacturing processes are rare. Furthermore, many older manufacturing engineers learned their profession on the job; they are able to understand well how present factories are run, but unless they receive further training, they will not know how to deal with the factory of the future. The problem is, there are people who know manufacturing, and there are those who know technology, but there are very few people who know both. Knowledge of both is required for the future factory, where manufacturing and technology go end effector-in-end effector. The demand for such people has already soared.

One of the reasons for this is that factories have not been very successful in attracting talented engineers to the factory floor. Many engineering graduates are afraid their talents and creativity will be stifled in factories, because they do not understand what kind of responsibilities manufacturing engineers will assume. Years ago, manufacturing engineers performed such banal tasks as helping maintenance people fix machines. Now, however, such engineers have responsibilities, such as running CAD/CAM computer simulations to test the effects of robotics implementation.

Japan has been more successful in attracting engineers to manufacturing, because their manufacturing engineers earn high wages and prestige. Also, since there are more than double the number of engineers proportional to the population than in the United States, there are more engineers to deal with in the first place. Furthermore, these manufacturing engineers are rewarded by promotions and often become plant managers. In the United States, manufacturing engineers are rarely promoted to manager. This situation seems to imply that plant managers don't need to understand the technology they use to be promoted and paid well. This can lead to various problems in the vital communication between managers and engineers.

One solution to this problem is to expose students to faculty who have experience with the modern industrial environment. Although many universities and colleges prefer to employ as faculty doctoral graduates with a strong research background, there is a growing awareness that practical experience or at least a familiarity with manufacturing processes is vital to a strong robotics education program. Educational institutions are also beginning strong applications programs through cooperative or internship programs. For example, some colleges require their engineering students to work part-time to balance their theoretical and academic studies with practical lessons.

Robots and Organized Labor

Labor unions have been formed in the past to protect the interests of people employed in many areas, particularly those in the secondary industries. Some of these organizations are quite large and powerful, and there has been some speculation that they would use their resources to halt or slow the implementation of robots in those industries that are labor union intensive, such as the autoworkers. However, this has not been the case to any great degree. For example, in one agreement between a labor union and a manufacturing company, the terms the labor union required for its members was a guarantee of a minimum of 60 days' notice before the installation and use of a robot or an automated manufacturing machine. This notice had to include a description of the device, the expected number of employees to be decreased as a result of the device, and the anticipated date of the start-up of the machine or robot. Why didn't the labor union leaders take a more aggressive posture toward the elimination of workers? Because they realize, as clearly as anyone else, that if our industries do not automate, they cannot compete with foreign and domestic industries that do, resulting in the collapse of many firms that now employ union workers. Some jobs are better than no jobs.

8.4 Resistance to Change

There will probably always be individuals or groups who resist change. For example, during the industrial revolution, a group of English workers called Luddites did their best to halt the progress of automation by smashing and sabotaging the new mills and looms and other machinery. They remind us that workers who are laid off because of changing technology usually take little comfort in the knowledge that the new machines will bring about better jobs for everyone in the future. Robots, as the automatic looms, will eventually bring about more and better jobs, but what about today? Banks and other lending institutions, upon which most workers must depend to buy and keep their homes and automobiles, are not noted for understanding attitudes toward the unemployed. It is easy to see why some individuals take a dim view of robotics, as well as other types of automated equipment, which cause changes in the work force. At the same time, industries must automate, or they will represent a serious underutilization of vital resources. The United States is no longer competing with itself. It has been matched and bypassed by other countries in a frantic production race. It is a race that, for many reasons, we simply cannot afford to lose. If we lose our manufacturing capabilities, for example, we face the risk of being unable to afford to make our own weapons. The political and social implications of being dependent on other nations for our own defense are quite obvious. Another reason is that we are currently one of the richest nations on earth. Our people enjoy one of the highest standards of living ever achieved by any society, any time. There are many nations who would keenly appreciate a chance to make our country dependent on them, as was demonstrated by the oil cartel countries in the 1970s. It is more than just a race for production. It is, ultimately, the same old struggle for survival. If we learn nothing else from history, it is imperative that we remember that those civilizations and countries that fail to adapt and adjust to new technologies wind up in last place in every race. One very good way to overcome the fears of people who believe robots will put them out of work is to communicate with them. Seeking them out and telling them what robots are, why they are used, and what they can and cannot do may do much to overcome their fears. If manufacturers demonstrate that they do not lay off workers solely as a result of using robots, then robots will cause no more consternation than any other piece of machinery.

As we have stated before, retraining is likely to be the single most important effect of robotization. The major problem anticipated is that some of the workers are not retrainable, largely because of a lack of motivation. The reasons for this lack can vary from poor self-image to outright fear of change. Many machinists and other workers whose jobs demanded physical prowess develop a certain hostility toward work that they consider somehow less "manly," and may be reluctant to retrain for jobs involving desk work or other nonphysical tasks (Cetron, 1983). It is probable that many current attitudes toward education, training, and technology must be reshaped to overcome these problems. For example, one suggestion has been to begin rewarding schoolchildren as publically and flamboyantly for mental achievements as they are now for physical achievements, since those physical achievements will probably not help the student earn a living after he or she graduates.

Other fears are that these new jobs will not offer many opportunities for advancement, will decrease interaction among workers, and will require much less supervision. As we mentioned before, one factor of a job that contributes highly to satisfaction is adequate supervision. Workers want to know there is someone to turn to when they run into problems they can't handle, someone to correct their mistakes, and someone to congratulate them when they do a good job. Humans are not like machines. They need the encouragement, attention, and emotional support of other humans in most working situations. Most of these fears can be easily eliminated where open communication and willingness to accommodate the requirements of both workers and management exist.

Another way to help overcome resistance to change and to promote easier adaptation to changing technology is to improve the quality of basic, public education. A recent report of a study performed by the National Commission on Excellence in Education reported that the state of public education in the United States was "deplorable" (Christiansen, 1983). The report indicated that our country is technically illiterate, largely because of the deterioration of basic skills, like reading, thinking, and studying. The College Board's Scholastic Aptitude Test scores declined steadily from 1963 to 1980. It reported that many high-school seniors are unable to read analytically or critically, rendering them unable either to draw inferences from reading material or to write persuasively. Too many students, the Commission report charged, do the minimum work required and take the least challenging courses when offered a choice. To remedy this situation, some 37 states initiated minimum competency tests that require basic literacy from their graduating students. Unfortunately, the effect of this requirement, according to the Commission, has been to make basic literacy the goal, rather than the origin, of learning, resulting in the lowering of educational standards for public schools as a whole.

Another area of trouble is mathematics and basic science knowledge. In one study, only a third of all high-school seniors could solve a mathematics problem that involved several steps. This situation has led to a rash of remedial math courses, with the result that nearly one-fourth of all math courses now offered in public schools are remedial in nature. This lack of competency in math is particularly discouraging when one compares the basic graduation requirements of other countries. For example, in the USSR, students are required to take 7 years of math; in Japan, the requirement is 6 years.

What has resulted appears to be a ridiculous paradox: our country seems to be charging full steam into a way of life steeped in and dependent on science and technology, but it seems to pull back as hard as it can on needed changes in public school education. There is a widening gap between the skills needed to thrive and prosper in our society and those skills offered by 12 years of public education.

For example, in the United States, the courses students take that have nothing to do with academics, such as food preparation or driver's training, are given the same credit toward graduation as physics or algebra, up to a certain limit. Although we may all agree that knowing how to drive and cook are certainly needed to function independently in our society, these are conceivably skills that can be learned at the student's convenience, during spare time or summer vacation. However, physics and calculus are not typically the sort of thing students want to learn on their own, particularly in their spare time. The differences between the preparedness of citizens from industrial nations that do insist on high academic achievement and that of our own is widening, and the consequences are more than a little conducive to panic among those who ponder the implications of these differences.

We can't have excellent teaching, of course, unless we have excellent teachers. There is currently little attraction for our best and brightest students to the teaching profession. The hours are long, the work is hard, and it is not as well paid as it might be. Math and science majors are often offered wages by industry that greatly exceed those offered to teachers. As long as teachers are considered unimportant enough to be paid low salaries, it is a wonder that so many do have the persistence, commitment, and idealism needed to be fine teachers. It seems clear that both the salaries and prestige of teachers must be increased to attract the best students to one of the most important jobs in our society.

Nothing happens easily. Poor pay, inadequate facilities, inadequate resources, and inadequate support of education can lead to a stagnant or declining society. Often, improvements call for higher taxes, which are usually politically unpopular. Yet, there is nothing that ignorance can accomplish except to breed more ignorance. If we continue to undersupport educational systems and are satisfied with high-school graduates who do not fit the job market, we must understand all the resulting implications and complications. We must also understand that our students are competing in an international contest that has been rigged in favor of advanced education with an emphasis on technology. It is a contest that we should not have to lose.

Questions

1. If you were called upon to help redesign a 4-year curriculum for high-school students to be implemented in the next 10 years, what are some of the courses you would recommend? Would you make these courses requirements for graduation or electives? Why or why not?

2. You are the owner of a company that makes ball-point pens. One of the factors leading to your success has been the wide variety of pens you offer. At present you employ 30 unskilled laborers to assemble these pens, along with 10 managers and supervisors. Now, suppose that you are offered a robot that can make pens in small batches with a higher degree of quality than your human laborers. Suppose also that the purchase of the robot would eliminate the need for all but two unskilled laborers and two managers and supervisors. This will lead to a significant savings over your present production costs. In 5 years, you calculate, your profits will enable you to expand production 150 percent. Would you buy the robot? If so, what would you do about the displaced workers? Would you feel they are the responsibility of the community and not of yourself? Why or why not?

3. What do you think might make the transition of labor from manufacturing to service easier for everyone? Consider government regulation, community ordinances, corporate policies, and labor union agreements.

Responsible Technology

9.1 Economic and Political Impact of Robots

The technologic expertise and capability represented by robots and associated technologies can be a political resource for various groups and nations. Many different groups have various ideas about how robots should be sold, implemented, utilized, and/or controlled. Some of these ideas have been the subject of congressional hearings by the Office of Technology Assessment and will undoubtedly be the subject of many future hearings and debates. Industrial robots are capital equipment; that is, they are machines purchased for the sole purpose of producing income for their owners. It is possible that robots may be capable of producing very large income for their owners. Individuals, groups of individuals, foreign interests, and the government are all present and potential owners of robots. Private ownership of robots by many individuals and companies is currently the most prevalent form of ownership in the United States. This kind of ownership carries certain responsibilities. In our modern nuclear age, it is not only our right but our responsibility to examine what kinds of impact modern technology is likely to have on society at large. For example, if robots have the potential to put many people out of work, some policy should be implemented from the very start to prevent this from happening. If there are safety risks associated with the use of robots in the workplace, then some guidelines should be worked out in advance, before anyone gets hurt.

The Subcommittee on Monetary and Fiscal Policy of the Joint Economic Committee (*Robotics and the Economy*, 1982) stated that robotics is expected to play an important role in the revitalization of our economy, that most of the possible negative effects of robotics might have been exaggerated, and that robotics should lead to improved working conditions, higher wages, and more jobs.

The Economic Recovery Tax Act of 1981 made it easier to exploit new forms of technology to their fullest potential. The Subcommittee's study indicated that robotics will provide us with a means to reverse or slow our productivity slowdown and start to grow again. Greater productivity growth means increases in wages and employment and better working conditions brought about by the elimination of human jobs in unsafe or undesirable environments.

The number of jobs taken by robots is likely to be less than the number of jobs that would be lost permanently if we did not match the efforts of overseas competitors in modernizing our industrial capacities. In many cases, there should be little or no direct job loss from introducing robots, but there will be substantial indirect gains in jobs arising from new kinds of work created in other sectors of the economy.

However, we need to be careful in our study of the employment effects of widespread robotics installations, so that public policies that are farsighted and thoughtful can be developed. Some of these policies will involve redirecting and expanding our educational and vocational training efforts to help provide the human skills and human capital needed to handle the new kinds of jobs that will be created for people as robots take over manual tasks. Also, greater cooperation that fosters growth and more efficient use of high technology is needed between research and private enterprise.

Robots rank with computers in their potential impact on labor, jobs, and quality of life and work. Even though the relative number of robots installed is still small, the continual technologic advances being made, along with improving investment and research incentives, make it very likely that we will experience tremendous growth in this new technologic tool within the next few years.

There is no inherent conflict between robots and humans. Productivity is not a struggle between humans and machines. The general public is perfectly capable of understanding what slow productivity growth means and how developments in robotics and automated manufacturing can help speed up that growth. We often seem to need reminding, however, that the entire goal of using the robots is to help, not hurt, humans. In other words, there is no evil plot of wicked scientists and capitalists to deprive hard-working citizens of their jobs and incomes. New machines are sometimes worrisome and often bring about tremendous changes in the way we earn a living. For example, the invention of the automobile practically eliminated an entire work category in the care and maintenance of horses. Many blacksmiths, farriers, and horse breeders were put out of work, but this did not happen overnight. It happened gradually over a period of about 20 years. As automobiles became more numerous and more affordable, people whose horses had died or become too old for service did not replace them with new horses, but with automobiles. Blacksmiths and farriers were certainly able to understand the great change in means of transportation, and the next generation learned to do other work. Too often, perhaps, the general public's ability to perceive and adjust to change is underestimated.

Of the small percentage of human workers that robots displace, only a tiny portion need face real unemployment. Robots can fill in the gaps left by natural worker attrition, such as retirement, disability, death, or relocation. There would be very few people that would face real unemployment, either as a direct or indirect consequence of robotics implementation. With thoughtful foreplanning, there need not be any.

Regulation of the Robotics Industry. Any time new technology is implemented, there is an adjustment period. People are generally mistrustful and cautious and want elaborate reassurance that the benefits will be worth the costs. One way to gain such reassurance is to increase government regulation of certain capital goods production. These restrictions might involve individual economic impact studies for each industry that wants to use robots, or the creation of an agency to oversee and control the number and type of robots sold. They might lead to laws that require hefty severance benefits for anyone who feels his or her unemployment is possibly related to the use of robots. Such policies, however, might also limit the freedom of robot users to the point where it might seem that they are being punished, rather than rewarded, for improving their productivity. This is where thoughtful policy planning comes into play. If workers whose jobs might be affected by robots are given firm assurance that they will not be shown the back door when a robot comes in the front, if robot manufacturers continue to specialize and differentiate so that there is healthy competition in all aspects, and if managers and planners carefully justify and plan for robot purchases, such fears can be minimized, and the need for government control may simply not exist. We have repeatedly emphasized the need for retraining workers whose jobs are affected by robotics. If an industry cannot afford to retrain its workers, it probably cannot afford to buy a robot, either. If the industry must roboticize to keep up with foreign competition, on the other hand, but cannot afford a retraining program, then there would be a need for some kind of government or educational or social program to provide assistance to the displaced workers (Vedder, 1982). We must keep in mind that we are talking about a small relative number of employees, perhaps 5 percent of all displaced workers.

Good public policy would include tax policies that encourage capital formation, as well as research and development. However, it would not need to dictate the allocation of resources between robots or other forms of capital investment. The place for government intervention, according to an Office of Technology Assessment report (*Exploratory Workshop on the Social Impacts of Robotics,* 1982), is in our educational systems, promoting retraining of workers facing temporary unemployment, in allocating research funds to colleges, technical schools, and universities, providing direct assistance to public vocational institutions, and educating the general public.

Robotics is expected to raise productivity, which will in turn bring material rewards to both employers and employees. New forms of employment will be created to offset jobs directly lost to robots, and working conditions and job safety will improve as robots assume dangerous or debilitating tasks. Robots will bring about qualitative as well as quantitative improvements in the goods and services produced in this country. This will allow our country to maintain and expand export markets now threatened by foreign competition. The expected boom in the robot market is essentially just a continuation of a mechanized labor trend that started centuries ago and has resulted in enormous material wealth and well-being.

9.2 Robot Development in Other Nations

Robotics is a worldwide development. England, Japan, West Germany, Sweden, France, and Russia are all engaging in vigorous robotics programs. For example, the Robotic Institute of America estimated that Japan had 14,000 installed, working robots by 1980 and some 32,000 by 1982. By contrast, we had less than half their robot population in both years. The Soviet Union's current 5-year plan includes the construction of 40,000 robots (Vedder, 1982). This is somewhat surprising, since the anticipated robots have neither electronic controls nor electric motors, which makes them less sophisticated than ours. The Swedish manufacturer, ASEA, has opened a plant in the United States and is vigorously pursuing implementation of robots in their own country, as well as exporting their robots to other countries.

Japan is our biggest competitor. The Japanese Industrial Robot Association expects that Japan will begin exporting 15 percent of its robots by 1985. The *Robot X News* (1983) reported that the value of robot sales in Japan was nearly triple those of the United States by 1983. The Japanese Industrial Robot Association reported an output of about almost 25,000 robot units valued at $618 million, and robot exports tripled to over $85 million. One Japanese manufacturer alone has a staff of 8000 engineers and scientists assigned solely to robotics research and development (*Robotics,* 1983). They are a people apparently committed wholeheartedly to robots and the idea that robots can be used to do just about anything.

The French Example. France is also engaging in the vigorous production and use of robots. France is the first country to have devised a formal public policy concerning robotics technology. Let's examine this policy to see just what it proposes to do. The Robotics Mission of the French Ministry of Research and Technology proposed a program to provide France with an internationally competitive robotics capability at all levels of design, manufacturing, distribution, and utilization (Stauffer, 1983). This program is a major national objective and had a projected cost of $350 million over a 3-year period. When the program started, France had an installed robot base of 850 robots, more than half of which were imported robots. They expect to have an installed base of 5000 robots by 1990. They are concentrating on six key areas of robotics: research and development, manufacturing, marketing, financing, socioeconomic, and education and training. The multidisciplinary nature of robotics, the French program designers believe, required the creation of an Interagency Robotics Committee (CIR) to coordinate the efforts between the electronic, data-processing, and machine tool industries and to determine priorities.

The goal is to have a French robotics industry within 5 to 10 years that is capable of a 25 to 30 percent productivity improvement in key sectors of the French manufacturing industry. To achieve these gains, they are concentrating in four major areas: strengthening vendor capability, increasing the utilization of robots in French industry, development of support services, and development of standardized products. Another goal is to cut manufacturing cost $1.5 billion, as well as create new areas for skilled employees.

The agency would further recommend which research efforts should have emphasis and advise what kind of robotics education should take place. Current projections include training some 100 people per year in general robotics, 500 people per year as industrial users of robots, 100 people per as specialists for research and robotics companies, and 1000 people over a 2-year period for the training of instructors and users of robots on a continuous basis.

The French Robotics Industry Association is proposed as the main organ for the dissemination of information, promotion, and publication of robotics applications. The principal research program that now exists is the Advanced Automation and Robotics Program. This is a team of 30 university researchers and 10 industrial engineers. They were projected to need $11.5 million for the next 3 years. Support for 300 research engineers, 60 scholarships over 2 years for doctoral candidates, $43 million in research contracts, and $10 million over 2 years for equipment is sought. The emphasis would be on industrial and manufacturing development, particularly in the areas of the role of robotics in improving quality, the installation and implementation of robots in manufacturing, the influence of robotics in developing new manufacturing processes, and the development of mobile robots for such uses as agricultural and materials handling. Other activities would include the identification of technical and industrial capabilities for system components, standardization of modular robotic components, research in flexible manufacturing systems, developing sources of key materials and components, and training robotics specialists. Testing and evaluation of pilot flexible manufacturing systems would be initiated. The funds for this development would come from major industrial companies that are already engaged in these activities. The funding recommended by the Robotics Mission would be limited to research and implementation studies. The French Robotics Industry Association and professional organizations, such as AFCET, ISF, and SEE, will be asked to develop a list of French system component suppliers and a detailed survey of research and development in robotics around the world and by major manufacturers, establish study groups for monitoring automation requirements of each manufacturing sector, encourage French manufacturers to participate in international exhibitions, and establish a high-quality robotics periodical for the dissemination of knowledge and research to manufacturers and users of robots.

Support would also be given to small businesses in the form of financial aid and technology transfer, as well as feasibility studies and economic justification studies. Other financial support would be given to users. The members of the Robotics Mission believe that robotics manufacturers will not make profits for a long period of time but robot users will see substantial savings within a very short time. Therefore, a robotics shared-savings plan that benefits both manufacturers and users of robotics would be initiated by establishing an industrial fund financed by a percentage of the cost savings generated by users. Users with limited capital funds would receive supplementary financing, and robotics manufacturers that can prove they are selling robots at unprofitable levels will receive supplementary financing, as well. This is expected to encourage robotics manufacturers to make robotics installations as productive and efficient as possible, to participate in the Industrial Mutual Fund support. Also, accelerated depreciation over robots over 2 years would be allowed until 1987. Financial programs that offer low-cost loans would also be made available to banks and financial institutions, both large and small, and users will be eligible for double the amount of support if installation is completed within a short time period of 6 to 9 months.

Special robotics institutes would also be created and given specialty area assignments, such as lightweight assembly robotics, robotic prototypes, the development of standards, and others.

It is not yet apparent how successful this public policy will be. We do know that France's productivity has recently risen dramatically in several areas, and their public policy may serve as a model for other nations that believe there is a need for such comprehensive control.

9.3 Growth of the Robotics Market in the United States

There have been many forecasts of the growth of the robotics market. Those of 10 years ago were overly optimistic, but it would appear that most of those today are overly pessimistic. One source estimates that annual robot shipments will exceed 2500 by next year. However, the National Bureau of Standards expects annual shipments to reach 4800 per year by 1985 and 17,100 per year by 1990. The RIA projects that robot production in the United States will be 24,000, with worldwide total of 78,000 units annually by 1990 (RIA Worldwide Survey, 1983). General Motors alone expects to spend some $1 billion by 1990 for 14,000 robots (Vedder, 1982). The estimated industrial expenditures (in millions of dollars) on robots from 1979 through 1985 is as follows (Obrutz, 1980, p. 49).

Industry	1979	1985
Electrical machines	16	164
Automative	15	54
Fabricated metals	16	67
Electronics	1.6	70
Heavy machinery	12	13
Others	19	71
Totals	79.6	439

There are three aspects of robotics growth that can have economic impact. The first is the magnitude of the growth, the second is the impact on unemployment, and the third is the impact on wages, profits, and prices. We will consider growth first. Robots can do higher quality work than humans, in many applications, because they are much more consistent. A human worker can get sleepy, bored, careless, tired, or distracted. A robot cannot. Qualitative improvements will increase reliability and the quantity demanded and permit high equilibrium, high market prices, and greater sales. This lowers the cost of producing a product because inspection and rejection costs fall.

When the cost per unit of robot-produced goods drops under the production costs from traditional production techniques, robots begin making profits. In calculating costs, the direct per unit capital costs of the robot, maintenance, indirect labor support, depreciation, additional property tax liability, and other costs are included; traditional costs include wages and salaries, fringe benefits, supervisory labor costs, and expenses associated with absenteeism, work stoppages, rejection, inspection, and so on.

While the cost of traditional labor-intensive techniques continues to rise, the cost of robotic-intensive techniques is falling. It is falling because technology advances, lowering the capital costs of robots per unit of output. In 1975, labor costs were much lower than they are today. Human wages rose, however, while the cost of robots continued to fall relative to prices in general. In the past couple of years, a robot welder has become cheaper than human welders, and that is why the robot can increase profits. The *Robot Times* (1983) quoted General Motors Chairman Roger Smith as saying that, "Every time the cost of labor goes up $1 an hour, 1,000 more robots become economical."

Public policies that raise the cost of robots per unit by requiring impact studies, elaborate justifications, or high severance costs can cause the cost of robots to rise, delaying the point at which robots become more profitable to employ than human labor. Also, taxes and depreciation rules that lower the marginal rate of return on capital can slow the introduction of new technology.

It has been reported (Kemle and Hall, 1983) that assembly robots is the fastest growing segment of the robot market. In 1982, five companies dominated the robotics industry and commanded 86 percent of all revenues. However, these companies now account for only about 59 percent of revenue, because there has been a trend toward fragmentation caused by the entry of many new companies into robotics. Of the five companies, three were public at the time of the report. There were at least fifty new robotics companies at the time of the report, divided into three main groups. The first group is composed of large corporations that sell robots purchased from other manufacturers with the intention of developing and selling their own hardware and software. These corporations have a lot of capital, but often lack technologic expertise and robotics experience. The second group is made up of small companies based in the United States that are diversifying their regular product line by adding robots. These companies often lack capital, electronics, and software expertise. The third group is composed of start-up companies with considerable technical expertise that are developing their own systems. Although some of the larger companies already have the necessary marketing and technical ability in electronics and software, most of the others are not so well situated.

Japan is expected to be the most significant foreign competitor. However, since the trend in robotics is toward a computer- and software-driven industry, the United States still maintains the lead in this area. With five large and at least fifty small manufacturers getting into robotics, it would appear that competition is growing faster than sales, and that the market cannot support so many manufacturers. There were $185 million in robot sales in 1982, but the start-up costs for a robotics manufacturer runs to $8 million. Therefore, the new companies are expected to experience negative cash flow until the market expands sufficiently to allow their support.

Venture capitalists are sometimes reluctant to invest in start-up companies until the industry "shakedown" has taken place, leaving only the most solid industries in place. This process has already begun to occur, with the acquisition of Unimation by Westinghouse and U.S. Robots by Square D as indicative of things to come. They expect little companies to be put out of business or bought out by bigger companies. The result of such reluctance is that many innovative, small companies are finding it difficult to come up with enough money to keep afloat until profits can be made.

Since many customers of robotics don't know how to integrate robots into useful systems, systems houses that provide robot users with needed information have sprung up all over the country. These businesses are generally quite small, founded and staffed mainly by engineers with robot applications expertise, but others have become quite large and substantial. These businesses are expected to keep growing, branching out, and specializing in different types of applications.

Manufacturers of programmable, high-functionality assembly robots fall into two categories: *Fortune* 500 companies with large resources that offer assembly robots with light payloads, designed either by their own engineers or imported from Japan or European countries, with enhanced electronics and software, or perhaps from a manufacturer they have bought out, and start-up companies with limited resources. These start-up companies either design their own robots or enhance imported ones. For example, GCA Corporation and Bendix design their own robots, but IBM and GE distribute Japanese robots. Westinghouse and Square D bought out domestic robotics manufacturers already in operation.

Another important area of significant economic impact is sensor systems. Kemle and Hall report that some $18 million was spent on 463 robot vision system units in 1983. This figure is expected to grow 42 percent annually through 1992. The reasons for this rapid growth rate are several. For example, entry barriers are low in peripheral systems manufacturing. Start-up costs for a vision system manufacturer are estimated at $5 million. There are currently 17 established manufacturers of robot vision systems. However, only a few are expected to survive the competition. The large companies that offer vision systems, for example, sell their general-purpose vision systems for stand-alone or robot-integrated systems for about $30,000. Application-specific vision systems can cost from $15,000 to $65,000.

The highest competition exists at the lowest technologic levels. The companies at the high-technology level already have large market shares because their products have already differentiated. However, there are high profit potentials in these high-technology companies because buyers are willing to pay large sums for the necessary applications software. For instance, one company had a reported 50 percent profit margin from its software alone.

These high profit margins will encourage many new companies in robotics and related businesses to spring up. We can expect some condensation and attrition to occur in these in adjusting to demand, but it is clear that the competition will be tough, which will most likely result in lower prices and high-quality systems. In short, it is clear that robotics and related industries are on their way to becoming Big Business.

9.4 Anticipated Benefits

Even though it appears that the first impact the installation of a robot has on a job is to reduce employment and create unemployment, this has not actually happened to any great extent. Because the robot has been introduced to improve productivity, the effect of the robot has been to make firms more competitive and able to increase benefits to its employees. The reduction in the work force resulting from the installation of robots can often be offset by the hiring of new maintenance personnel, workers to manufacture robots, and computer programmers to instruct the robot. If the firm takes a positive attitude toward displaced employee retraining, these new workers will be the old workers, only retrained. What will change is the structure of employment. This structural change can greatly benefit human workers. As the robots replace workers in monotonous, physically tiring, and mentally unrewarding tasks, human workers will be placed in more challenging, more interesting, and more rewarding jobs.

The safety factors are also tremendous. Breathing paint fumes, asbestos fibers, and coal dust are just a few of the many things humans now do with full knowledge that their health will be adversely affected. For instance, coal miners take their jobs knowing that they will be screened regularly for pneumoconiosis, a disease that results from prolonged exposure to coal dust. When their x-rays show that they are in the first stages of the disease, they are retired from the mines and given compensation. This is, as you can imagine, a very costly way to go about mining coal, both monetarily and in human sacrifice. If robots can replace these workers, we could not only prolong many lives, but could apply the money intended for their eventual compensation to their immediate retraining. Meanwhile, the new generation of workers who are now in school could be educated so well that working in mines is never one of the alternatives.

One argument we have encountered is that our society needs garbage collectors and janitors and cannot use a generation that is all college educated, because educated people don't want to be garbage collectors or janitors. Robots are about to change all this. We will have robots to do the dirty work, and there will be no further excuse not to educate and train humans to their full capabilities. No human can possibly work physically harder than a robot.

The new jobs that are created by robotics are usually skilled, challenging, untiring, and safe. Robots will raise the productivity of the human workers. When marginal productivity, which is the increased output, rises, it becomes more profitable to hire more workers. The changes in supply should produce greater demand, which will stimulate employment. This greater productivity should also raise the living standards for everyone connected with the user firm because of the robot's impact on wages, profits, and prices. The prices are expected to change in the following manner. If a factory sells its product for $50, and could hire an additional human worker on the production line, this would increase the revenue by roughly twice that amount, or $100. If a robot, instead of a human, is added and one additional worker is hired to maintain the robot, and the robot is capable of producing five times the number of products as a human could have produced, the additional output of the firm would increase from $100 to $250. It thus would be profitable to hire more workers at any given wage, whereas before, it was only profitable to hire workers at under $100. However, the increased willingness of a firm to supply output because of declining production costs resulting from the higher worker productivity will force it and competing firms to lower prices to sell their increased output. Wages will go up, and the consumers of the products will benefit. Their dollars will be able to buy more goods from that firm. What should result is that everyone will gain through the adoption of the new technology—workers from higher wages, consumers by lower prices, and the firm by higher profits.

If many workers lose their jobs, however, from the introduction of the new technology, profits would not increase because there would be decreasing demand for the increasing production. This is why good planning is crucial. There is no profit if demand does not rise along with supply, and there can be no rising demand with rising unemployment.

The 4 to 7 percent of American skilled workers whose jobs could be overtaken by robots in the next 10 years are not apt to be shown the back door for another reason. It has been shown that no one is better able to train and supervise a robot for a given job than the human who used to do that job. After all, he or she is familiar with all the unexpected and rare occurrences that happen on the job from time to time and will best be able to help the robot to deal with it or advise the programmer how to adjust the robot.

9.5 Alternatives

As Americans, we have the freedom to choose to actively encourage or discourage the introduction of robotics in our society. All current economic analyses indicate that the price of not roboticizing will be higher unemployment, poorer economic conditions, dropping out of the production race as a manufacturing nation, and becoming an entirely service-oriented society. There are many who believe that our society should become totally information oriented, and that our future prosperity lies in becoming suppliers of information, not goods. That would be fine, but there does seem to be a catch. If we lose our manufacturing capabilities, we may eventually be forced to depend on other countries for defense weapons and capital-producing equipment. Perhaps we would have to shop for our defense from France or the USSR. Furthermore, without our own manufacturing systems, our own robotics experts would have to work for foreign companies. The fact is, as long as there exist other countries that do use robots, the United States will have to, too, or else be prepared to face the consequences. If the Japanese offer robot-produced cars and televisions at a lower price and superior quality, American dollars will go to Japan. The money we spend on American goods is usually recirculated in our own economy. The money we spend on foreign goods is not necessarily ever reintroduced into our own country. Yet, we cannot be expected to pay high prices for inferior goods simply because they are American made. For one thing, we enjoy considerable consumer freedom and protection from such policies, and for another, we don't know anyone who can resist a bargain. The government could introduce policies that make foreign-made goods more expensive by levying high tariffs and penalties, but these policies would be unpopular with both our allies and our consumers, who benefit from the competition brought about by foreign goods. Black markets would probably become inevitable, undoing the efforts of the restrictions. There are some things you can't easily make Americans do, and paying more than they have to for goods is one of them. It didn't take long for the automobile industry to figure this out. Not surprising, therefore, was its decision to introduce robots in efforts to increase quality and lower costs to stay competitive.

When the assembly line and mass production of the automobile were first introduced into our country, the number of hours required to produce a car dropped 56 percent. Sales increased 10 times, resulting in nearly quadrupling the number of human workers needed to maintain and service the cars.

Many of the industries that have roboticized were already experiencing loss of employees. These industries were forced to lay off workers because demand for their products fell. Why did they fall? Competition from foreign countries was one factor, particularly from Japan. The Japanese had higher productivity and were able to sell their goods at a lower cost. Installing robots did not necessarily save all these industries, but it did give them a fighting chance to keep competing on a more nearly equal level. If you were going to hold a contest between two people, say, to see who could produce the greatest number of cars in a given week, would you say it was fair if one of the parties had two robots to help but the other had only one? Japan is the competitor with two or more robots to our one or less, overall. In the automotive industry, the ratio of robots is much closer. However, you can see the inequity in production that will result until we can match or surpass the Japanese capability.

9.6 Conclusions

Historically, breakthroughs in labor-saving technology have increased output, wages, and employment and reduced inflation. They also accelerate the rate of economic growth that results in material benefits for everyone. There are secondary benefits, as well. Some of these, in relation to robots, are reducing the amount of danger and monotony in industry and improving the quality of consumer goods.

The largest productivity gains are often made through the introduction of the new technology made possible by educational institutions. However, new technology has a habit of becoming old in a very short time. More changes in technology have occurred in the past 20 years than occurred in the previous 50,000. This rapid rate of change rather obviously has many ramifications. Politicians and government leaders frequently do not fully understand all the aspects of the technology for which they must develop guidelines—no one can be expected to entirely understand or foresee the significance of new technology. It is ultimately up to individual users, manufacturers, and beneficiaries of robots to take those positive steps to ensure that this new important technology is introduced in a socially and politically responsible manner.

Questions

1. Public policies could either raise or lower the cost of robots. Discuss the policies which could effect such changes.
2. How can robots be introduced into a workplace without laying off present workers?
3. What is the goal of the French public policy concerning robotics technology? Do you think this is a good policy? Why or why not?
4. How might the political structure of competing manufacturing nations change with the introduction of millions of working robots?
5. We mentioned that the United States could not lose its manufacturing capabilities without severe ramifications. What are some of these, and do you agree?

Robots Today and Tomorrow

10.1 Robot Developments

Now that we have a good understanding of robots and what they can and cannot easily do and how they have been applied in the past, let's look at some of the research taking place today. In this way, we will gain some insight to future applications and uses of robots. As we will see, the possible applications are nearly unlimited, offering the opportunity to today's students to become involved with exciting, important developments.

Automated Factories

One of the bigger impacts robots will have in the near future is on factory and warehouse design. Using sensor systems, automatic transfer systems, and hierarchical control, entire processes can be automated, eliminating the need for people in unsafe or tiring factory jobs. Several factories are already completely automated, such as the Mazak machine tool plant in Florence, Kentucky. At this plant, the entire operation is handled by mobile and stationary robots and other automated equipment. Other roboticized factories are now being built in Japan to assemble robots.

Eventually, most manufacturing plants will probably be totally automated. This will come about as robot systems become more accurate, dynamic performance is improved, and further developments are made in sensor systems, hierarchical control, and knowledge-based management systems.

The Need for Standardization

Little standardization exists among the various manufacturers and producers of robot peripherals, controls, sensors, and other components. This means that turnkey robotic systems are few and far between, yet this is what most robot buyers want.

There are several reasons for the lack of standardization. One is that each robot producer naturally wants to be the sole source of a particular type of system. Another is that many robot users want to implement their robots in conjunction with unique and sometimes old automated equipment, which calls for special design or installation. Perhaps the most important reason is that there is not yet any "best" robot or robot design. The lack of standardization is, however, frustrating to many people. It helps keep costs high, installation lengthy, and overall familiarity with robotics minimal. For example, industry experts have fought long and bitterly over the very definition of the word *robot*. Other terms, including many we have used in this book, are also hotly contested. We put much consideration into the phrase "end effector," and found the word *manipulator* in the literature to refer to anything from the working end of a robot to the robot itself. Lack of standardization also makes the job of properly training a robot technician difficult, because the teacher or trainer must gain familiarity with as many different types of robot systems and units as there are companies that produce them. The solution to standardizing lies in a spirit of cooperation developing among engineers, designers, robot producers, and potential robot users. Although competition among robot manufacturers may tend to preclude such cooperation, we believe that, as robot demand increases, so will the degree of standardization among various robots. Also, as old equipment and automated machinery wear out, new equipment already designed to be integrated with robots will replace them. Such integration capability will call for further standardization.

With standardization and the increased availability of turnkey systems for robots developing, the robot industry is likely to expand dramatically. The capabilities provided by sensory devices being developed all over the country, coupled with the advances in computer capabilities, will result in an ever-increasing list of new applications and unexpected benefits. Intelligent, mobile robots are on their way and will be used everywhere—in space exploration, medical care, agriculture, and homes. Let's look at some of the research now underway to see where the next generation of robots is headed. As we do so, it should become clear that the potential applications of robots are limited only by our ability to imagine them.

10.2 Ongoing Research and Future Applications

Advances in Robot Intelligence

One of the biggest differences in the robots of the future will be in their intelligence (Hall, 1983). When robots have sensors that can sense and respond to certain things in their environment, they will be able to do many more things than they can now. They will appear almost human in the execution of some tasks. Although many of the tasks that robots now perform in industry do not require sensors, the presence of sensors cannot help but improve that performance. For example, modern spray-painting robots cannot recognize whether there is actually a part in front of them to be sprayed. That is a function of the total automated process, or the conveyor belt. However, if such a robot is equipped with a sensory device, such as a photoelectric diode, it can recognize gaps in the production line. Another example is in arc welding. Robots that do this welding cannot tell in advance exactly where the line of join is to be made. Also, burr removal in cleaning castings is an application in which, because the burrs appear at random, the robot can greatly benefit from a tactile sensory device so that it can "feel" the burrs and remove them. Applications that would be heavily dependent on sensory feedback would include the clothing, footware, and rubber industries. Such robots with an extension arm could be used for processing and maintenance, by the military, in homes, or in difficult and dangerous environments.

The Military

The military has always been on the forefront of new technology. This is no accident, of course. Many modern societies contribute a very significant portion of their available resources to the various branches of the military. With these enormous resources, a good deal of time, talent, and money can be spent on the research and development of constructive, as well as destructive, machines and methods. In the United States, the military is very interested in the development of robots. Some of the reasons given for this are that there is at present a downward trend in both quantity and quality of recruits and retention is falling, and yet the complexity of tasks, processes, and machinery used by the military is increasing. Robots may be able to fill many gaps left by unqualified or unavailable recruits. Examples of the kind of applications being considered are mapping, dangerous materials handling, submarine tracking, and battlefield surveillance.

Domestic and Entertainment Robots

One of the most thought-provoking (and fun) applications of intelligent robots is their use in or around the home. Wolkomir (1983) discusses several possible applications, as follows:

Window washer
Coffee server
For delivering drinks
Security patrol
Leaf raker
Cat feeder
Bathtub cleaner
Bath giver
To store and retrieve household items
To read a child stories
Appliance repairer
Hair stylist
Caretakers for invalids
Pets

Friend to play checkers or chess
Companion

Although some items on this list are a bit fanciful, some are quite feasible and may possibly be implemented within the next 10 years, despite the anticipated high cost and minimal efficiency of a robot that could perform such tasks. One reason for this is that our modern lifestyles leave us so little time for leisure that many of us are willing to pay premium prices for machines able to assume a significant portion of time-consuming but boring home maintenance tasks. Many people in business are becoming aware of the potential sales in the area of domestic robots. For example, suppose that a labor-saving device becomes available for $30. Now, imagine that 30 million Americans buy that device. This would create total sales of nearly $1 billion per year. A potential billion-dollar market is very attractive! Perhaps this explains why some of the best robot researchers are busy developing domestic robots.

For example, Joseph F. Engelberger, one of the best known experts in robots, has a robot named Isaac (in honor of Isaac Asimov) that is an articulated arm on wheels. In one test at his office, Engelberger had Isaac open a cabinet, pull out a mug, pour coffee into it, and ring a bell to signal that the coffee was ready. Cute, you say, but does it do windows? Bill Bakaleinikoff of Superior Robotics claims to have one that does just that, using magnets attached to the edges of the windows for the robot washer to follow.

Robots as companions might seem a silly idea at first. However, a robot companion could be an extremely useful and recreational tool for elderly, disabled, or handicapped people. For example, a mobile robot with a long arm and gripper could be handy for retrieving and storing items on otherwise inaccessible shelves. A robot sentry system that could monitor the home for fire, undesired entry, or gas leaks, for example, and automatically alert rescue or police units would free many elderly or home-bound disabled people from much worry. Such a sentry robot was developed by Bart Everett while he was a student at the Naval Postgraduate School in Monterey, California. His robot, called ROBART, can patrol the home and warn the household of smoke, fire, gas, intruders, flood, earthquake, and storms. It also contains a computer-generated vocabulary of 300 words. This robot is shown in Figure 10–1. Companion robots that could play and record tape or play games would do much to relieve the boredom or solitude of many people. Japanese scientists are reportedly performing serious research into the problem of developing robots to care for and entertain the elderly and disabled.

Outdoor maintenance robots are also being seriously considered. One of the authors is currently helping develop a prototype lawn-mowing robot for both industrial and domestic use and predicts that one could be commercially available in only a few years.

There are already several "personal" robots on the U.S. market. One of these is Brains on Board, called BOB, for short. It is 3 feet tall and uses infrared sensors attuned to the frequency of human body heat to find and follow people in a room. It is, of course, mobile, can avoid obstacles and pivot, and comes equipped with two microprocessors with 3 megabytes of memory. It also has a voice synthesizer. It is built by Androbot, of Sunnyvale, California (Roessing, 1983). When its battery runs low, it is programmed to recharge itself. It can be programmed to say or sing anything you like. Another

Figure 10–1. ROBART is a sentry robot built by Bart Everett. It is designed to function completely on its own, with a navigational system to avoid colliding with obstacles, and a battery monitor that allows it to recharge itself as needed. (Courtesy of Dr. Bart Everett, Springfield, Virginia.)

home-use robot is the DC-2, built by Southern California Android Amusement Company. It was one of these robots that friends bought for *Playboy* publisher Hugh Hefner in 1980. He uses the robot to greet guests, serve drinks, and entertain. His particular model comes with a color television, a videotape recorder, a color camera, and a drink tray. A similar robot, OPD2, is probably the only official robot policeman. It was purchased in 1981 by the Orlando, Florida, Police Department for public relations and educational use. A demonstration android called Robot Redford created headlines when it delivered the commencement address at Anne Arundel Community College in Maryland. It was created by Bill Bakaleinkoff of Superior Robotics in Petaluma, California. It currently earns $1200 per day at trade shows and fairs. It attracts crowds by singing, dancing, and making humorous remarks.

Entertainment robots are usually simply remote-control androids used to attract and entertain crowds. One popular robot is Denby, the ProMotion Robot, built by World of Robots Corporation and shown in Figure 10–2. Entertainment robots often find unexpected uses. One robot that has been used recently as a therapeutic tool for mentally handicapped and autistic children is SICO, a 6-foot-plus android built by International Robotics of New York City. Roessing (1983) reported that this robot, on a recent flight from New York to Los Angeles, paid for its own ticket with a credit card, ordered a meal, and, when asked if it preferred smoking or nonsmoking seating, responded with, "I only smoke transistors." This robot has been programmed to speak several languages and to display emotional cues. Bakaleinikoff has also built a Robot Louis Stevenson android

Figure 10–2. Denby, the ProMotion Robot, can be rented from the World of Robots Corporation in Jackson, Michigan, for entertainment or promotional use. It can rotate and nod its head, move its shoulders, grasp with its hands, and has full mobility with remote voice. (Courtesy of World of Robots Corporation, Jackson, Michigan.)

and Stereobot, a mobile sound system for party use. He also produces Robot Wars, a game that involves two robots that fight each other with electric ray guns.

Neiman-Marcus offered a home-use robot in its 1981 Christmas catalog. This robot, called ComRo I, was designed by Jerome Hamlin of Comro, Inc., New York. It was advertised as able to open doors, walk the dog, take out trash, water plants, and sweep floors. The cost of this robot was about $15,000. In 1983, *The Sharper Image* catalog listed a robot with an vacuuming arm attachment for about $3500. Named RB5X (Arbie for short), it is built by RB Robot Company of Golden, Colorado. It was designed to be used for home and educational use and contains its own microprocessor. These robots represent a big technical step from the toy remote-controlled robots of previous times. Other robots that are currently being offered are HERO I, by Heath Company and ComRo TOT by Comro. These and the RB5X feature on-board computers and sensing capabilities that enable them to talk, move, determine distances, sense light, grip objects, and teach.

The producers of these various types of robots expect the home-use robot market to expand at about the same rate as home computers. Robots that can be controlled by home computers are likely to be the first robots that find a wide market, simply because such robots are less costly. However, as the example of the home robot that went from $15,000 to $3500 in 3 years shows, robots are going to become more affordable by the public. Since people are generally quite fascinated by robots, price is likely to be the main barrier to widespread use. As soon as the price of a home robot drops under $1000, they will very probably become as common in homes as computers.

One technical problem that must be surmounted with the current domestic robot batteries is that a fast-charging battery is needed. Today's batteries discharge and charge at about the same rate. A desirable battery would discharge at a slow rate but charge at a fast rate. It would probably be tolerable for the robot to take a "lunch break," but certainly not a 4-hour break on the day you plan to have guests. Another alternative would be switchable battery packs. However, the simplicity required for practicality might then be gone. Technology in this area has progressed a great deal with the developments in electric cars. However, further improvements are needed.

Robots in Medical and Patient Care Applications

An important development in robotics is their use in medical care as extensions in fulfilling the insufficient abilities of handicapped or disabled people. Japanese researchers report that they are trying to develop robots that could be used as guides for the blind. Their advantage over a Seeing Eye dog would be that robots need not be fed or cared for so diligently as a living creature. Also, it would create no dirt or waste products that are often troublesome for the elderly to dispose of. The Japanese have developed a two-legged robot named Wabot I, completed in 1973, although they continue to improve on this design. Although this robot is designed for self-control, such robots could be used as mobile extensions for immobile people.

Also in Japan, a robot for assisting bedridden patients has already been developed. It consists of an integrated control apparatus, a television camera and monitor, a command device, an articulated robot, a convenience rack, and an automatic transport vehicle. The convenience rack is like a bookshelf or open cabinet, which holds various items, such as food, water, or books, the patient might need or use in a given day. The items on this rack are displayed on the TV monitor. The patient may select an item or items from the menu, which the transport vehicle moves to the convenience rack. The articulated robot, which is a smaller version of the large industrial arm, but with nine joints, loads the selected items. The vehicle then moves back to the patient's bedside, and the robot assists the patient to perform given tasks, such as holding a book or handling the patient's food or drink. If this robot is successfully implemented, it will represent a major breakthrough in patient care. Nurses will be free to perform the more complicated and sensitive tasks for which they have been educated and trained. After all, fetching items for patients does not require a college degree.

Robots in Remote or Hostile Environments

One of the most exciting capabilities robots may provide in the future is that of being our eyes and hands in places that are otherwise inaccessible or hazardous to us, such as outer space, the ocean floor, deep mines, and dangerous environments, such as burning or gas-filled buildings. Underwater robots could be used to build or repair ocean structures, such as offshore oil rigs, to gather samples from the ocean floor, to drill, or in search-and-rescue operations. These robots would not be so limited by size as their above-water counterparts and could be powered by water or air jets. Current research in this area is being performed mainly in Japan.

Another environment open to robots is space. We have already seen the development of the space shuttle arm and Viking Lander. The space shuttle manipulator is shown at work in Figure 10–3. Such manipulators may be the forerunners of robots that can build space platforms and factories or manufacture products that cannot be efficiently made on Earth. Several researchers predict that robots will be used in such applications, as well as in Moon and planet exploration and mining, maintaining orbiting space stations, repairing structures or space vehicles, and even in manufacturing more robots (Dorf, 1983).

Japanese researchers are working on a search-and-rescue robot for emergency use in burning or gas-filled buildings. This robot would incorporate most of the elements of the most sophisticated robots—visual and speech feedback systems, mobility, and sensitive manipulators. It would also have self-protecting systems, such as a sprinkler to keep itself cool, and could manipulate a water hose or chemical extinguisher.

Many other hazardous environments mentioned earlier, such as coal mines, radioactive areas, and chemically polluted environments, along with arctic and desert areas, are all potential breeding grounds of the next generation of robots. Continued support in research and development will result in overcoming the last barriers to our exploration of the world around us.

Robots in Education

One very exciting application of robotics is in the field of education. Robot literacy is a natural extension of computer literacy; after all, a robot is simply a computer that moves.

Figure 10–3. The Space Shuttle Challenger's Remote Manipulator Arm grasps the Shuttle Pallet Satellite (SPAS-01) during proximity operations on June 22, 1983. The scene has within it a few reflections on the window through which it was photographed. SPAS-01 was developed by the West German firm Messerschmitt-Boelkow-GmbH (MBB). (Courtesy of NASA.)

The success of computer languages such as LOGO with small children is that they quickly learn that the robot "turtle" shown on the screen can be moved up, down, left, and right to define a very interesting animation sequence. Our children are getting to be skilled robot operators before they put their second roll of quarters into the video games. The skill of controlling the motion of something is what robotics is all about. What better place to learn this than on a computer terminal in a simulated mode. The same procedure is used to train airline pilots. They "fly" a simulator before they take off in the real thing. An interesting thing to note is that the attention span of a child, which is usually considered very short, can extend into hours when tackling a challenging video game. The applications of robots to education in elementary school has not been seriously considered, perhaps because of the cost factor. The ability to perform the same task over and over is as natural for a machine as it is unnatural for a human, yet children often need this repetition for learning. There is also an important ego factor difference between being told we are wrong by a machine and by another human. The machine corrects our logic but does not challenge our self-image. We believe robots will not be far behind the computers in elementary schools. Today, there are few people who can claim 20 years' experience with robots, and these few are making a big impact on our society. Imagine what might be accomplished if our entire younger generation enters the work force with this many years of experience.

Robots in middle and high schools is another area that has barely been explored. Students with 7 or 8 years of experience with computers and robots are going to be ready and able not only to use robots but also to build them. At a recent science fair, three of the major entries were self-built robots by high-school students. If students never have to ask why they need to learn math, algebra, trigonometry, analytic geometry, or calculus because they are using them in their latest robot creations, then something very wonderful will have happened.

Robots are being seriously considered in technical schools and colleges not only because of the job opportunities but also because of the obvious need for this technology in our nation. This book is but one example of the attempts to disperse information about this important technologic area. With the advent of the educational robots, such as Microbot's Minimover and HERO I, which are in the $1000 to $2000 price range, technical schools can acquire laboratory equipment to provide hands-on experience with robots. A HERO I robot is shown in Figure 10–4. This popular robot has several sensors, a speech synthesizer, an arm with a gripper, a teach pendant, and microprocessors. Assembling this robot is an educational process, and the uses are unending. Various programs can be developed to use the HERO as a sentry, a servant, or just to let it wander through obstacles on its own. A small servo-driven robot called the Rhino and its larger relatives may be safely and easily used in an educational environment to study the use of industrial robots. Entire work cells or automated factory concepts are demonstrable for a small fraction of the cost of an actual cell or factory. Many universities have benefited from gifts of modern robot equipment directly from the manufacturers or through government grants.

HERO I is currently being used in many introductory robotics and advanced engineering courses. It comes in a kit, which the students must assemble. It is a self-contained, electromechanical android that can be programmed to speak any language, pick up small objects, travel over predetermined paths, and repeat specific functions on a programmed schedule. TOPO, another educational android, is being distributed for use with Apple computers. The RB5X is different because it is equipped with its own memory and programs and sonar and tactile sensors that permit it to learn from experience. Its hardware and electronics can be altered or added to. It was made available to children at certain computer camps around the country in the summer of 1983.

Robots in Agriculture

Robots are being developed for agricultural purposes in the United States and other countries around the world. One motivation for developing robotics in this area is the high cost and seasonal nature of the labor. Often, labor comprises a major proportion of

Figure 10–4. HERO I, Heathkit's education robot, is made available to students as a kit. The students shown constructed the robot as part of their robotics studies at the University of Cincinnati.

the cost of producing the agricultural product. This sometimes makes produce vulnerable to crushing competition from imported produce from countries that have low labor costs.

At the First International Conference on Robots in Agriculture held recently at the University of Florida (Isaacs, 1983), researchers from around the world reported their work in agricultural robot applications. They ranged from automatic rice field harvesting to sheep shearing to self-guided tractors and combines. Most prerobotics work in agriculture has been in the area of feedback control systems in the equipment. However, with sensors and microprocessors, intelligent robot systems can be used in applications requiring selective harvesting or dealing with live animals. For example, researchers at the Agricultural Engineering Department at the University of Florida at Gainesville are investigating a robot citrus harvester. Such a robot is shown in Figure 10–5. It not only must be mobile and have a versatile manipulator, but it needs a visual sensor to differentiate between ripe and unripe fruit. Also, the location and orientation of the trees and fruit are quite variable, requiring complex computer programming to enable the

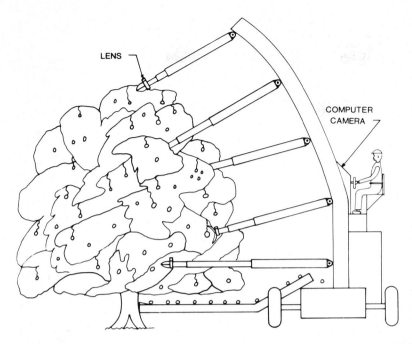

Figure 10–5. Illustration of the University of Florida researcher's citrus harvesting robot system. (Courtesy of Dr. G. E. Coppock, Florida Department of Citrus, University of Florida, Gainesville.)

robot to recognize an orange tree, find the fruit, and harvest only the desired fruits. The development and testing of the multiple manipulator prototype will take place over the next few years. It is anticipated that the current high cost of labor required for orange harvesting makes the cost of such a robot economically as well as technically feasible.

Intelligent combines and tractors are another development in agricultural robotics. The concern motivating these robots is that it is extremely difficult for humans to guide farm equipment in very consistent, even rows. However, particularly in an application such as cultivation or harvesting, it is desirable that no variation occurs between rows to maximize harvest. Automatic guidance systems have been tested that allow less than a few millimeters variance (Harrell, 1983).

One of the most spectacular and advanced examples of an agricultural robot system is the sheep shearer developed by a team of researchers at the University of Western Australia and shown in Figure 10–6 (Trevelyan, et al, 1983). The system incorporates a sensor that keeps the cutting head just above the animal's skin, accommodating variances due to the animal's breathing and other factors, and results in a very low injury rate. This development is particularly exciting to many people because it is the first known application of a robot that performs tasks on a living creature. Who knows? Perhaps there will soon be robot barbers for people as well.

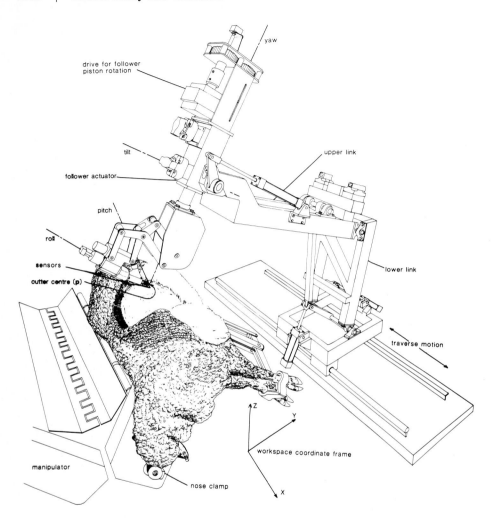

Figure 10–6. Australian scientists at the University of Western Australia have developed a robot sheep shearer. (Courtesy of Dr. Stewart J. Key, University of Western Australia, Department of Mechanical Engineering.)

Mobile Robots

Mobile robots that cannot only perform work tasks but also keep track of their location and navigate around obstacles are being studied in many research centers, such as Stanford University, Carnegie-Mellon University, and the University of Cincinnati. The industrial robot cart, which can transport parts and materials, is currently one commercial application. Many other applications can be expected to reach a commercial stage in the next few years.

Walking robots are an important special class of mobile robot that have recently walked out of research laboratories, such as the one at Ohio State University, into commercial application. Odetics, Inc., which has headquarters in Anaheim, California, introduced the Odetics six-legged walking robot in 1983. This machine cannot only walk, but was demonstrated climbing into a pickup truck, climbing down, then lifting up and carrying the truck. Odetics expects their robot to be used in many applications, such as cargo handling and storage, irrigation, harvesting, field inspection, mining applications, such as extraction and roof control, nuclear reactor applications, such as inspection and emergency tasks, and in the military. The ODEX I is shown in Figure 10–7.

10.3 Conclusion

For centuries, one of the favorite themes in stories and literature is that of a genie or magic machine that will do whatever its owner commands. We have all probably wondered what we would do if we were to find an Aladdin's lamp. Would we waste its power on foolish whims, or would we be wise enough to use the magic to accomplish something good and lasting? In many ways, the robot promises to be this generation's magic lamp. It has the potential to give us riches, power, leisure, and ease. Whether

Figure 10–7. The ODEX I, a six-legged walking robot, has many possible uses, particularly in environments with rough terrain. ODEX I is the first walking machine ever devised that can lift and carry loads in multiples of its own weight. In a demonstration, the first functionoid, weighing only 370 pounds, lifted the back end of a 2200-pound truck. Walking in a normal profile, it is in the process of turning the truck's position 90 degrees. (Courtesy of Odetics, Inc., Anaheim, California.)

it actually does, however, is up to how wisely and well we use it. The ultimate goal of the use of robots always should be to help, not hurt us. This calls for thoughtful planning, intelligent policies, and foresighted decisions on the parts of leaders in fields that will use robots, from factory owners to medical researchers. We are approaching a new era in our continuing industrial evolution. It can mean freedom from all mindless work, freedom to develop our uniquely human abilities to their fullest extent, and freedom to explore all the frontiers that we could never physically explore ourselves. Continued sensitivity, however, to the effects of technology on all people is of paramount importance for the wisest and happiest implementation of this new technology.

Questions

1. What are some of the problems that can be attributed to the lack of standardization in the robotics industry?

2. What are some possible applications for both industrial and nonindustrial robots that were not mentioned in this chapter? What kinds of problems might one encounter in these applications?

3. What opportunities do you think robots will open up for people?

4. In your opinion, in what area of robotics is the most significant research taking place?

5. Overall, do you think robots will be mostly beneficial? Why or why not?

References

"AL User's Manual," Stanford AI Memo AIM-323, Stanford, CA.
Albus, J. S., *Brains, Behavior and Robotics*. Peterborough, NH: BYTE/McGraw Hill, 1981.
Anbe, Y., Y. Takahashi, Y. Arimura, et al., "A Computer Controlled Robot Cart," *Proceedings of the Second International Symposium on Industrial Robots*, IIT Research Institute, Chicago, IL, May 1972, pp. 115–126.
Arnold, P., and P. White, *The Automotive Age*. Pound Ridge, NY: Holiday House, 1963.
Asimov, I., *The Complete Robot*, New York: Doubleday, 1982.
A Survey of Industrial Robots. Dallas, TX: Leading Edge Pub., Inc., 1982.
Awerman, A., and D. Cappello, "Positive Employee Relations Paves the Way for Robots," *Robotics Today*, Vol. 4, No. 6, Dec. 1982, pp. 35–36.
Ayres, R. U., and S. M. Miller, *Robotics, Applications and Social Implications*, Cambridge, Mass.: Ballinger, 1983.
Barbera, A. J., *An Architecture for a Robot Hierarchical Control System*, NBS Publication SP-500-23, Dec. 1977.
Berg, N. S., *1983–84 Directory of North American Robotics Education and Training Institutions*, Robotics International, Society of Manufacturing Engineers, Dearborn, MI, 1983.
Bonner, N. S., and K. G. Shin, "A Comparative Study of Robot Languages," *Computer*, Dec. 1982, pp. 82–96.
Boorstin, D. J., *The Discoverers*, New York: Random House, 1983.
Buffington, P. W., "Making (Body) Sense," *Sky*, November 1983, pp. 85–87.
Burstall, A. F., *A History of Mechanical Engineering*, Cambridge, MA: MIT Press, 1965.
Carrico, L. R., "Robotic Training—A Comprehensive Plan," *Proceedings of the Sixth Robots Conference*, March 1982.
Cetron, M. J., "Jobs With A Future," *Job Forecasting*, 98th Congress Report No. 6, April 1983, pp. 20–43.
Christiansen, D., "Spectral Lines," *IEEE Spectrum*, Vol. 20, No. 10, Oct. 1983.
Christiansen, H. N., "MOVIE. BYU Documentation," Brigham Young University, Dept. of Civil Engineering, Utah, 1978.
Cincinnati Milacron, "Robots. A Manager's Guide," Cincinnati Milacron Marketing Co., 1982.
Coiffet, P., *Robot Technology, Vol. 1, Modeling and Control* (English Trans.), Englewood, N.J.: Prentice-Hall, 1983.
Davey, P. G., "Intelligent Robots in the 1980's," Application of Microprocessors in Devices for Instrumentation and Automatic Control, 1983.

Davies, O., *The Omni Book of Computers and Robots,* New York: Kensington, 1983.

Dawson, B. L., "Moving Line Application With a Computer Controlled Robot," SME Paper MS77-742.

DeCosta, F., *How to Build Your Own Working Robot Pet,* Blue Ridge Summit, Penn.: Tab Books, Inc., 1979.

Dorf, R. C., *Robotics and Automated Manufacturing.* Reston, Va.: Reston Publishing Co., Inc., 1983.

Engelberger, J. F., *Robotics in Practice,* New York: American Management Assoc., 1980.

Ernst, H. A., "A Computer-Operated Mechanical Hand," Sc.D. Thesis, Massachusetts Institute of Technology, Cambridge, Mass., 1961.

Everett, H. R., "A Microprocessor Controlled Autonomous Sentry Robot," Naval Postgraduate School Thesis, Monterey, Calif., 1982.

Exploratory Workshop on the Social Impacts of Robotics, Congress of the United States, Office of Technology Assessment, Washington, D.C., February 1982.

Fitzgerald, C. T., "Robotics In Engineering Technology," *Proceedings of the American Society for Engineering Education,* Rochester Institute of Technology, Vol. 2, L. P. Grayson and J. M. Biedenback, eds., June 1983, pp. 723–728.

Fu, K. S., Ed., "Robots and Automation," *Computer,* Dec. 1982.

Fukui, I., "TV Image Processing to Determine the Position of a Robot Vehicle," *Pattern Recognition,* Vol. 14, Nos. 1–6, 1981, pp. 101–109.

Geduld, H. M., and R. Gottesman, Eds., *Robots Robots Robots,* Boston, Mass.: New York Graphic Society, 1978.

Gibbons, J. H., Ed., "Exploratory Workshop on the Social Impacts of Robotics," Congress of the United States, Office of Technology Assessment, 1981.

Glorioso, R. M., and F. C. Colon Osorio, *Engineering Intelligent Systems.* Bedford, Mass.: Digital Press, 1980.

Goertz, R. C., "Manipulators Used For Handling Radioactive Materials," *Human Factors in Technology,* E. M. Bennett, Ed., New York: McGraw-Hill, 1963.

Gore, A., *Job Forecasting,* Committee on Science and Technology. Washington, D.C.: U.S. Government Printing Office, 1983.

Green, R. H., "Welding With Traversing Robots: A New Concept in Line Tracking," SME Paper MS79-791.

Groover, M. P., *Automation, Production Systems, and Computer Aided Manufacturing,* Englewood Cliffs, NJ: Prentice Hall, 1980.

Guterl, F., "An Unanswered Question: Automation's Effect on Society," *IEEE Spectrum,* May 1983.

Hall, E. L., *Computer Image Processing and Recognition.* New York: Academic, 1979.

Hall, E. L., "Getting a Grip on Robotics," *The Bent of Tau Beta Pi,* Vol. LXXIV, No. 4, Fall 1983, pp. 31–35.

Hall, E. L., J. Lesac, W. E. Woodrick, and D. J. Soukup, "An Educational Computer Vision and Robotics System," *Proceedings of the IEEE Computer Society Conference on Pattern Recognition and Image Processing,* Las Vegas, Nev., June 1982, pp. 647–749.

Hall, E. L., J. B. K. Tio, C. A. McPherson, F. A. Sadjadi, "Measuring Curved Surfaces for Robot Vision," *Computer,* December 1982, pp. 42–54.

Harmon, L. D., "Automated Tactile Sensing," *International Journal of Robotics Research,* Vol. 1, No. 2, Summer 1982.

Harrell, R. C., "Application of Intelligent Robots to Agriculture," *Proceedings of the First International Conference on Robotics in Agriculture,* Tampa, FL, 1983.

Heer, E., Ed., *Proceedings of the First National Conference on Remotely Manned Systems,* California Institute of Technology, Pasadena, 1973.
Heiserman, D. L., *Build Your Own Working Robot.* Blue Ridge Summit, Penn.: Tab Books, Inc., 1976.
Heiserman, D. L., *How To Build Your Own Self-Programming Robot.* Blue Ridge Summit, Penn.: Tab Books, Inc., 1979.
Heiserman, D. L., *Robot Intelligence With Experiments,* Blue Ridge Summit, PA: Tab Books, Inc., 1981.
Hemenway, J., "Start with BASIC Investigations to Understand Robotics, AI," *Engineering Design News,* May 12, 1983, pp. 169–177.
Hohn, R. E., "Application Flexibility of a Computer Controlled Industrial Robot," SME Technical Paper MR-76-603, 1976.
Holland, S. W., L. Rossol, and M. R. Ward, "CONSIGHT-I: A Vision Controlled Robot System for Transferring Parts from Belt Conveyors," *Computer Vision and Sensor Based Robots,* G. G. Dodd and L. Rossol, Eds., New York: Plenum Press, 1979, pp. 81–100.
Holmes, J. G., "Justifying a Robot Machining Center," Adapted from "An Automated Robot Machining System," presented at the 9th International Symposium and Exposition on Industrial Robots, Robotics Institute of America and Society of Manufacturing Engineers, March 13–15, 1979, Washington, D.C.
Holt, H. R., "Robot Decision Making," *Proceedings of the Second Robots Conference,* Autofact 1, Second National Industrial Robots Conference and Exposition, RIA and SME, Detroit, Mich., Nov. 1–3, 1977.
Howard, J. M., "Focus on the Human Factors in Applying Robotic Systems," *Robotics Today,* Vol. 4, No. 6, Dec. 1982.
Huston, R. L., and F. A. Kelly, "The Development of Equations of Motion of Single-Arm Robots," *IEEE Trans. on Systems, Man, and Cybernetics,* Vol. SMC-12, No. 3, May/June 1982, pp. 259–266.
Isaacs, G. W., "Conference Wrap-Up," *Proceedings of the First International Conference on Robotics in Agriculture,* Tampa, Fla., 1983.
Job Forecasting, Hearings Before the Committee on Science and Technology, U.S. House of Representatives, 98th Congress, Washington, D.C., April 1983.
Johnson, K., and D. Hanify, Eds., *Proceedings of the First National Symposium on Industrial Robots,* April 2 and 3, 1970, IIT Research Institute, Chicago, Ill.
Johnson, K., and D. Hanify, Eds., *Proceedings of the Second International Symposium on Industrial Robots,* May 16–18, 1972, IIT Research Institute, Chicago, Ill.
Kato, I., and Y. Hasegawa, "State of the Art in Robotics,"
Kemle, J., and C. Hall, Report E3-7, Harvard University Graduate School of Business Administration, Boston, Mass., 1983.
Kutcher, M., "Automating It All," *IEEE Spectrum,* May 1983, pp. 40–43.
Loofbourrow, T., *How To Build A Computer Controlled Robot,* Rochelle Park, N.J.: Hayden Book Co., 1978.
Lundstrom, G., B. Glemme, and B. W. Rooks, *Industrial Robots—Gripper Review,* Kempston, Bedford, England: International Fluidics Services, Ltd., Nov. 1977.
Machine Vision Systems, A Summary and Forecast, TechTran Corp., Naperville, Ill., 1983.
Malone, R., *The Robot Book,* New York: Push-Pin Press, 1978.
Mattox, J., "Robots Don't Just Handle Things—They Do Things Too," *Automation,* Nov. 1976.
MAZAK, State of the Art Technology in Manufacturing Systems, (Company Brochure), Yamazaki Machinery Works, Ltd., Florence, Ky.

McLean, C., M. Mitchell, and E. Barkmeyer, "A Computer Architecture for Small Batch Manufacturing," *IEEE Spectrum,* May 1983, pp. 59–64.

Merritt, H., *Hydraulic Control Systems,* New York: John Wiley & Sons, 1967.

Miller, E., "Technology and Automation: The Nissan Experience," *Survey of Business,* Summer 1982, pp. 26–30.

Minimover-5 User Reference and Application Manual. Menlo Park, Calif.: Microbot, Inc.

"Mobile Robot Accelerates Paint Spraying Operation," *Robotics Today,* June 1982.

Moravec, H., "The CMU Rover," *Proceedings of the National Conference on Artificial Intelligence,* Pittsburgh, Penn., August 1982, pp. 377–380.

Mujtaba, S., and R. Goldman, "AL User's Manual," Stanford Artificial Intelligence Laboratory Memo AIM-323, 1979.

Munson, J. H., "The SRI Intelligent Automation Program," *Proceedings of the First National Symposium on Industrial Robots,* IIT Research Institute, Chicago, Ill., April 1970, pp. 113–117.

Naisbitt, J., *Megatrends,* Warner Books, New York, 1982.

Nitzan, D., "Assessment of Robotic Sensors," *Proceedings of the First International Conference on Robot Vision and Sensory Controls,* Stratford-upon-Avon, England, April 1–3, 1981, pp. 1–11.

Obrutz, J. J., "Robots Swing Into the Arms Race," *Iron Age,* July 21, 1980.

Obrutz, J. J., "Robotics Extends a Helping Hand," *Iron Age,* March 19, 1982, pp. 59–83.

Office of Technology Assessment, Congress of the United States, "Report on the Economic Impact of Robotics," 1982.

Operating Manual for the Cincinnati Milacron T3 Industrial Robot, Version 3.0, Robot Control With Restructured Software, Cincinnati Milacron, Industrial Robot Div., Lebanon, Ohio.

Paul, R. P., *Robot Manipulators.* Cambridge, Mass.: MIT Press, 1981.

Proceedings of the First International Conference on Robot Vision and Sensory Controls, April 1–3, 1981, Stratford-Upon-Avon, England, IFS (Conferences) Ltd., Oxford, England: Cotswold Press, Ltd.

Proceedings of the Robotics and Remote Handling in Hostile Environments, National Topical Meeting, April 1984, Gatlinburg, Tenn.

Raphael, B., *The Thinking Computer, Mind Inside Matter,* San Francisco: W. H. Freeman and Company, 1976, pp. 275–288.

Rebman, J., and K. A. Morris, "A Tactile Sensor with Electrooptical Transduction," *Proceedings of the Symposium on Intelligent Robots: Third International Conference on Robot Vision and Sensory Controls,* Cambridge, Mass., Nov. 6–10, 1983.

Reich, F. R., et al., "Robot to Solve Rubik's Cube," SAE Technical Paper 830341, SAE International Congress and Exposition, Detroit, Mich., Feb. 1983.

Richardson, I., *Social Problems.* New York: Random House, 1980.

Robertson, G. I., "Robotics Networks: Keys to the Automated Factory," *Systems and Software,* Oct. 1982, pp. 53–55.

Robertson, I., *Social Problems,* 2nd Ed., New York: Random House, 1980.

Robotics Institute of America, *RIA Robotics Glossary,* P.O. Box 1366, Dearborn, Mich.: Robotics Institute of America, 1984.

Robotics, Report on the Hearings of the Committee on Science and Technology. Washington, D.C.: U.S. Government Printing Office, 1983.

Robotics Application Guide, General Dynamics Corp.

Robotics and the Economy, Subcommittee on Monetary and Fiscal Policy of the Joint Economic Congress of the United States. Washington, D.C.: U.S. Government Printing Office, 1982.

Robotics Institute of America, *1982 Directory,* Dearborn, Mich.: RIA, 1982.
Robotics Institute of America, *Worldwide Robotics Survey and Directory,* Dearborn, Mich.: RIA, 1983.
Robotics . . . Start Simple, MACK Corp., Flagstaff, AZ, 1982.
"Robots Add Flexibility to Small Parts Assembly," *Robotics World,* June 1983, pp. 16–20.
Robot Times, Published by the Robotics Institute of America (RIA), P.O. Box 1366, Dearborn, MI, 1983.
Robot X News, Vol. 2, No. 5, Oct. 1983.
Roessing, W., "Brave New Robots," *SKY,* Aug. 1983, pp. 12–21.
Safford, E. L., *Handbook of Advanced Robotics,* Blue Ridge Summit, Penn.: Tab Books, Inc., 1982.
Sagan, C., *Broca's Brain,* New York: Random House, 1979.
Science and Invention Encyclopedia, Vol. 2, p. 162., H. S. Stuttman Co., Inc., New York, 1977.
Seim, T. A., "Applying Microprocessors to Machine Tool Design," *Computer Design* (Part 1, p. 86, March 1980; Part 2, p. 90, April 1980).
Shunk, D. L., K. Oestreich, and E. A. Long, "Applying the Systems Approach and Group Technology to a Robotic Cell," *Robotics Today,* Vol. 4, No. 6, Dec. 1982.
Sitkins, F. Z., "Survey of Robotics Education and Industry Expectations," in *Proceedings of the 1983 Annual Conferences of the American Society for Engineering Education,* Rochester Institute of Technology, L. P. Grayson and J. M. Biedenback, Eds., June 1983, pp. 716–722.
Society of Manufacturing Engineers, *Directory of Robotics Education and Training Institutions in Colleges, Universities and Technical Institutes,* 1 SME Dr., P.O. Box 930, Dearborn, Mich., April 1984.
Stackhouse, T., "A New Concept in Robot Wrist Flexibility," *Proceedings of the Ninth International Symposium on Industrial Robots,* March 1979.
Stauffer, R. N., "Developing the Robot Workplace," *Robotics Today,* Vol. 4, No. 6, Dec. 1982.
Stauffer, R. N., "France Maps Robotics Strategy," *Robotics Today,* Vol. 5, No. 2, April 1983, pp. 72–73.
Susnjara, Ken, *A Manager's Guide to Industrial Robots,* Corinthian Press, Shaker Heights, Ohio, 1982.
Taylor, R. H., P. D. Summers, and J. M. Meyer, "AML: A Manufacturing Language," Int. J. Robotics Research, Vol. 1, No. 3, Fall 1982, pp. 19–41.
Tanner, W. R., Ed., *Industrial Robots,* Vol. 1, *Fundamentals,* Dearborn, Mich.: Society of Manufacturing Engineers, 1979.
Tanner, W. R., Ed., *Industrial Robots,* Vol. 2, *Applications,* Dearborn, Mich.: Society of Manufacturing Engineers, 1979.
Tarvin, R. L., "An Off-Line Programming Approach," *Robotics Today,* Summer 1981.
Tarvin, R. L., *Course Notes for Robotics I,* Department of Mechanical and Industrial Engineering, University of Cincinnati, 1983.
The OMNI Future Almanac, R. Weil, Ed., New York: World Almanac Pub., Newspaper Enterprises Assoc., Inc., 1982.
"The Report of the French Robotics Mission," English translation by Lanshaw and Co., Freestone, Calif., 1982.
The Sharper Image Catalog, The Sharper Image, 755 Davis St., San Francisco, Calif., 1983, p. 63.
Trevelyan, J. P., P. D. Kovesi, and M. C. H. Ong, "Motion Control for a Sheep Shearing Robot," *Proceedings of the First International Symposium of Robotics Research,* Massachusetts Institute of Technology, Cambridge, Mass., M.I.T. Press, 1983.

Truxal, C., "The Professionals: Experts are in Short Supply," *IEEE Spectrum,* May 1983.

Unimation Application Notes 15 and 19.

"User's Guide to VAL," Unimation Publication No. 398H2A, Westinghouse, Unimation Div., Danbury, Conn.

Vedder, R. K., "Robotics and the Economy," Report of the Joint Economic Committee, Congress of the United States, Washington, D.C., March 1982.

Weisel, W. E., "Career Opportunities in Engineering Technology," *Proceedings of the Thirteenth ISIR-7,* Vol. 2, Detroit, Mich.: Robotics International of the Society of Manufacturing Engineers, April 1983, pp. 19-1 to 19-9.

Wellborn, S., "Machines That Think," *U.S. News and World Report,* Dec. 5, 1983, pp. 59–62.

Whitney, D. E., and J. L. Nevins, "What is the Remote Center Compliance (RCC) and What Can it Do?" Presented at the Ninth ISIR, March 1979.

Wolkomir, R., "Robots At Home," *OMNI,* April 1983, pp. 70–76.

Wright, J., in *Job Forecasting,* Committee on Science and Technology, Washington, D.C.: U.S. Government Printing Office, 1983.

Glossary

Accuracy: A measurement of the ability of a robot manipulator to go to a specified commanded point in space.

Actuator: A device in robots that converts electric, hydraulic, or pneumatic energy into motion, as in a linear actuator, servo motor, or rotary actuator.

Adaptability: The ability of a device to alter or modify its program or function in response to changes in its environment without human intervention. A robot that could shut itself off when it sensed the presence of a human in its work envelope would be adaptive.

Adaptive control: A method of control in which actions are continuously adjusted in response to feedback.

Address: A name, label, or number that is given to a register, memory location, or device.

Air motor: A pneumatic device that converts air pressure to mechanical force. Many servo mechanisms use compressed air to power motion, particularly in explosive environments where an electrical spark could be dangerous.

Algorithm: A given, detailed set of rules, usually in the form of mathematical equations, designed to achieve a specified result in a finite number of steps.

Alphanumeric: A term applied to characters that are either numerals, letters of the alphabet, or special symbols. Such characters are called alphanumeric to distinguish them from characters, special graphics, or machine language symbols, for example, that are used to command a computer. User-friendly systems use alphanumeric symbols because they are more easily understood. These alphanumeric symbols are translated into machine command symbols by a special language program. BASIC is such a language. A command given in BASIC is converted by a special program into machine symbols that instruct the computer to perform a certain operation.

Analog: The representation of measurable quantities by means of continuous physical variables, such as translation, rotation, voltage, or resistance.

Analog-to-digital: Often referred to as A/D, this refers to the conversion of continuous qualities to digital quantities.

And: This is a command used in computer languages, usually referring to the logical operation that takes on the value 1 only when both input variables have the value 1; otherwise, it has the value 0.
Android: A machine that resembles a human in physical appearance, particularly as pertains to entertainment and fictitious robots.
Anthropomorphic: Having a human shape; an anthropomorphic robot is one with rotary joints that enable it to perform tasks in a manner resembling that of a human; a jointed-arm robot.
Architecture: The physical and/or logical structure of a computer program or manufacturing process.
Arm: That part of the robot manipulator comprised of an interconnected series of mechanical links and joints that support and move the wrist and end effector through space.
Array: A set of storage locations in a computer memory that may be addressed using an index variable, for example, A(I); I = 1, 2, 3, 4, 5.
Articulated: The property of being jointed or connected in such a fashion that the joined parts are movable.
Artificial intelligence: This term refers to the capability of a computer to perform operations that simulate human intelligence. Examples of such operations are learning, adaptation, recognition, classification, reasoning, self-correction, and improvement.
ASCII: American Standard Code for Information Interchange, an 8-bit code used to represent alphanumeric, punctuation, and special characters for use in control.
Assembler: A computer program that translates mnemonic operators, such as ADD, into machine code numbers.
Assembly language: A computer language in which each command can be translated into a machine language command.
Asynchronous: Events, functions, or operations that do not occur simultaneously or in a specifically timed manner.
Automaton: A machine resembling a live creature in physical appearance and/or movement; especially referring to the mechanical dolls, animals, and statues of the seventeenth through nineteenth centuries that operated on cams or clockwork mechanisms to perform given operations in a lifelike manner.
Axis: A path of travel in the rotary or translational, prismatic (sliding) joint in a robot; a degree of freedom.
Binary: Referring to the property of having only two values. A binary number system, for example, has only two numerals, 0 and 1. Combinations of these two numerals are used to describe all information. A binary picture has only two colors, black and white.
BPS: Bits per second, the rate of transmission of binary information; also called baud.
Branch: An instruction to a computer that is usually given to cause the control unit to obtain the next instruction from a location other than the next sequential one. Branches may be of two types, conditional or unconditional.
Break point: The point in a computer program at which the program may be interrupted by external intervention.

BSC: Binary synchronous communication refers to the communication line that controls the exchange of digital data between computers or terminals across telephone lines, in which signals are timed with a clock signal.

Bus: A parallel channel along which data can be sent.

Byte: A byte is a group of 8 bits, each of which is a symbol representing binary digits that command a computer.

Calling sequence: A basic set of instructions used to start, initialize, or transfer control to and from a subroutine.

Cartesian coordinates: Also called rectangular coordinates, this refers to a set of three numbers that define the location of a point in a rectilinear coordinate system, composed of three perpendicular axes, and referred to as x, y, z coordinates. A Cartesian coordinate robot's manipulator moves in these three directions only.

Character: A symbol that expresses information to its reader. Computer characters are those that form a code that converts all information into binary digits.

Chip: A chip is an integrated circuit. All computers and computerized mechanisms contain these chips on which the instructions for the operation of the mechanism are performed.

Clamp: A function of a pneumatic robot gripper that controls the grasping and releasing of an object.

Clear: A computer command that usually directs the computer to replace information in its storage by zeros.

Closed loop: This is a robot control system that uses feedback to control its operation. In a closed-loop system, the robot control uses a feedback loop to measure and compare the actual system performance to the programmed desired performance and then makes adjustments accordingly.

Code: A set of computer instructions to perform a given operation or solve a given problem. Codes can be either symbolic, in which case the instructions are written in nonmachine language, or nonsymbolic, which means the instructions are written in machine language. The latter, although much harder to learn to read and write, is much faster than the former.

Compiler: A compiler is a computer language that translates symbolic operation codes into machine operation codes in a one-to-many manner. The machine language resulting from the compiler is the translated and expanded version of the original. It is much more easier to use than assembly language.

Complex sensors: Vision, sonar, and tactile array sensors that enable a robot to interact with its environment.

Computer: A device that uses information to perform prescribed operations and supply results of those processes.

Computer-aided design: Also referred to as CAD, this is the use of a computer to assist in creating or modifying design parameters.

Computer-aided manufacturing: Also called CAM, this refers to the use of a computer in the management, control, and operation of manufacturing.

Conditional jump: A jump subject to the result of a comparison made during the program.

Contact sensor: A device that detects the presence of an object, or that measures the amount of force or torque applied by the object, through physical contact with it.

Continuous path: A method of controlling a robot in which the commands or input specify all points in the desired path of motion.

Contouring: Controlling the path of a robot manipulator between successive three-dimensional positions or points in space.

Controlled path: A method of robot motion control in which intermediate points between command points are interpolated by the robot controller. This allows the robot to move in a straight line between programmed points.

Controller: A robot's brain, a computer-instructed system that directs the motion of the robot end effector, such as a microcomputer or some other programmable device.

Control unit: That portion of a computer that directs the automatic operation of the computer, inteprets computer instructions, and initiates the proper signals to the other computer circuits to execute instructions.

Convolution: A generally applicable local operation that can be used to perform such operations as noise filtering, edge extraction, and contour following.

Counter: A device or memory location that can be set to an initial number and increased or decreased by an arbitrary number.

CPS: Characters per second.

Cycle: A sequence of operations that is repeated regularly; also, the time it takes for a robot to run through its programmed motions.

Cylindrical coordinates: Spatial coordinates defined by two distances and an angle. A cylindrical coordinate robot's manipulator can move in one angular and two linear directions.

Data: A collection of facts or alphanumeric characters that are processed or produced by a computer.

Data link: Equipment that permits the transmission of information in data format.

Data processing: Any procedure for collecting data and producing a specific result.

Debugging: The process of determining the correctness of a computer routine, locating any errors, and correcting such errors; also, the detection and correction of malfunctions of the computer itself.

Deductive capability: The ability to draw conclusions from knowledge, rules, and general principles.

Degree of freedom: A motion variable for a robot axis, usually referring to a rotation or extension.

Diagnostic routine: A test program used to detect hardware malfunctions in a computer and/or its peripheral equipment.

Differential positioning: The difference in positions obtained by providing pulses of compressed air to the air motor in opposite directions, resulting in more accurate positioning.

Digit: One of the n symbols of integral value ranging from 0 to $n - 1$, inclusive, in a scale of numbering of base n. One of the 10 decimal digits, 0–10.

Digital control: The use of a digital computer to perform processing and control tasks in a manner that is more easily changed than an analog control system.

Digital image: A numerical representation of a picture seen by a TV camera, which is used as a robot's "eye." The robot's "brain" analyzes this digital image to enable the robot to recognize an object.

Digital image analysis: A multistage process that leads to a computer's "understanding" a digital image, the recognition of given objects, or the recognition of certain attributes in given objects in the image. Stages of this process may be image digitizing, image processing, feature extraction, and pattern recognition.

Digital-to-analog converter: Also referred to as D/A, a device that transforms digital data into analog data.

Discrete variable: A variable that takes on only a finite number of values.

Disk memory: A programmable, bulk-storage, random-access memory consisting of a magnetizable coating on one or both sides of a rotating thin, circular plate.

DNC: Direct numerical control. A system in which a digital computer is directly connected to one of more numerically controlled machine tools and controls the machining operations.

Documentation: A group of techniques used to organize, present, and communicate recorded specialized knowledge.

Edit: A computer mode that allows the creation or alteration of a program. This mode is available with the robot arm passive.

Encoder: A transducer used to convert position data into electrical signals. The robot system uses an incremental optical encoder to provide position feedback for each joint. Velocity data are computed from the encoder signals and used as an additional feedback signal to assure servo stability.

End effector: The tool attached to the end of a robot wrist that actually performs the work; also, gripper or process tooling.

Engineering units: Units of measure as applied to a process variable, such as pounds per square inch (psi) or degrees Fahrenheit.

EPROM: Erasable programmable read-only memory. A nonvolatile memory used to store a program.

Error: The difference in value between a given response and the desired response in the performance of a controlled machine, system, or process.

Executive control program: A main system program designed to establish priorities and to process and control other programs.

Feature: A certain characteristic of an image, such as an edge, contour, or silhouette or transitions from black to white or vice versa.

Feature extraction: The process of eliminating undesired information from a computer image and extracting useful information.

Feedback: The signal or data fed back to a command unit from a controlled machine or process that denotes its response to the command signal; also, the signal representing the difference between the actual response and the desired response, which is used by the commanding unit to improve the performance of the controlled machine or process.

File maintenance: The processing of a master file to handle nonperiodic changes in it, such as the changing of the name of a file, copying, or backing up the file.

Filter: A device used to suppress certain noise in a signal or image that interferes with the desired features.

Fixture: A device needed to hold a workpiece in the proper position for the performance of work.

Flag: A bit used to store 1 bit of information. A flag has two stable states and is the software analogy to a switch.

Flexible automation: This term refers to the multitask capability of robots; also, multipurpose, adaptable, or reprogrammable machines.

Flexible manufacturing: Production with machines that are capable of making a different product without retooling or any similar changeover. Flexible manufacturing is usually carried out with numerically controlled machine tools, robots, and conveyors under the control of a central computer.

Flip-flop: A bistable device (capable of assuming two stable states), it may assume a given stable state, depending upon the pulse history of one or more input points and one or more output points. The device is capable of storing 1 bit of information, controlling gates, and other functions; also, a toggle.

Flowchart: A graphic representation of a sequence of steps or operations using symbols that represent the operations.

Format: The predetermined arrangement of characters, fields, lines, page numbers, punctuation marks, and other elements that refer to input, output, and files.

FORTRAN: Formula translator. The language for a scientific procedural programming system.

Fusing: The operation of joining separate materials, as in welding.

Gap: An interval of space or time associated with an area of data-processing activity (record) to indicate or signal the end of that record.

Geometric processing: The process of taking measurements characteristic of the geometry of certain objects in an image. Examples are area, orientation, perimeter, number, and location of circles.

Global: Overall, as in a global measurement of an object.

Global coordinate system: A reference coordinate system in a fixed location; contrasted with a joint coordinate system, which moves with the robot joint.

Global operation: The transformation of the gray-scale value of picture elements according to the gray-scale values of all the elements of the picture. Threshold operations are examples of global operations.

Gradient: A vector indicating the change of gray-scale values in a certain neighborhood of a pixel. The gradient can be obtained by applying a difference operation on the neighborhood.

Graphic system: A system that collects, uses, and presents information in pictorial form.

Gripper: An end effector of a robot, usually referred to as a "hand." A gripper is designed to pick up, hold, and/or release the part or object being handled; also referred to as a manipulator.

Hard copy: Any form of computer-produced printed document.

Hardware: The mechanical, magnetic, electrical, and electronic devices of which a computer or robot is built.

Hexadecimal: A notation of numbers in the base 16 with characters 0, 1, 2, 3, 4, 5, 6, 7, 8, 9, A, B, C, D, E, and F.

High-level language: A simplified computer programming language that uses English-like statements for instructions and is oriented to the program to be solved or the procedure to be used.

Histogram: The relative frequency of gray-level values in a digital image.

IC: Integrated circuit. A solid-state, microcircuit contained in a chip of semiconductor material, usually silicon.

Image: A spatial array of information or picture, such as that presented by a TV camera.

Image enhancement: The process of improving the quality of the appearance of an image by use of such operations as noise filtering, contrast sharpening, or edge enhancement.

Image processor: A device or program that selects and interprets data to determine an object's position, location, shape, and size.

Initialize: A program or hardware circuit that returns a program, system, or hardware device to its original state.

Input: The data supplied to a computer for processing; also, the device used to accomplish this transfer of data.

Instruction: A set of bits that causes a computer to perform a prescribed operation. A computer instruction consists of an operation code that specifies the operation to be performed, one or more operands, and one or more modifiers that modify the operand or its addressee.

Intelligent robot: A robot that chooses between actions according to the way it senses its environment.

Interface: The concept of interconnection between two objects with different functions.

Interrupt: A break in the normal flow of a system or program that allows an action to be performed and the flow to be resumed from that point at a later time.

I/O: Input/output.

Joint: A single degree of arm rotation or translation.

Joint-interpolated motion: A method of coordinating the movement of joints in such a way that all joints arrive at the desired location simultaneously.

K: An abbreviation of the quantity 1000, generally used as a measurement of memory capacity. For example, a memory with a capacity of 1K means the memory has a capacity of 1024 words, or the number 2 to the tenth power.

Knowledge engineering: The use of artificial intelligence techniques and a base of information or knowledge about a specific activity used to control systems automatically. This type of system is called as "knowledge-based system."

Label: An ordered set of characters used to identify an instruction, program, quantity, or data area.

Labeling: The process of assigning different numbers to the picture elements (pixels) of different blobs in a binary image.

Language: A defined group of representative characters or symbols combined with specific rules necessary for their interpretation. The rules enable an assembler or compiler to translate the characters into forms meaningful to a machine, a system, or a process.

Lead through: Programming a robot by physically guiding the robot through the desired actions.

Limit switch: A switch actuated either by some part or motion of a robot or machine to alter the electrical circuit.

Limited sequence: A simple or nonservo type of robot with movement controlled by a series of limit or stop switches. Also called a "Bang Bang" robot.

Linkage: A means of communicating information from one routine to another.

Load capacity: The weight that a robot can manipulate with a fully extended arm. Most robots can handle heavier loads when full arm extension is not required. Gross load refers to the total weight that must be lifted by the robot, including the weight of its own end effector. Net load refers to the weight of a load less the weight of containers, tooling, and the end effector. (*See also* Rated load capacity.)

Loader: A program that operates on input devices to transfer information from off-line to on-line memory.

Local operation: Transformers of the gray-scale value of the pixels according to the gray-scale values of the element itself and its neighbors in a given neighborhood. Examples are gradient, sharpening, smoothing, and edge extraction operations.

Location: The storage position in memory uniquely specified by a given address.

Log: A record of values and/or an action for a given function.

Loop: The repeated execution of a series of instructions for a variable number of times.

LPM: Lines per minute.

LSI: Large-scale integration. High-density integration of circuits for complex logic functions. LSI circuits can range up to several thousand logic elements on a silicon chip a tenth of an inch square.

Machine intelligence: The design of machines that incorporate the ability to apply knowledge from a data base or sensors to permit it to manipulate its environment.

Machine language: A language written in a series of bits, which are understandable by a computer. The "first-level" computer language, compared with a "second-level" assembly language or a "third-level" compiler language.

Machine vision: The capability of a robot to "see" conferred on it by a vision system.

Macro: A source language instruction from which many machine-language instructions can be generated.

Magnetic disk storage: A storage device or system consisting of magnetically coated metal disks.

Mainframe computer: The principal computer in a system or network of computers.

Manipulator: That part of the robot that performs mechanical movements.

Manual control: A device containing controls that manipulate the robot arm and allow for the recording of locations and programming motion instructions.

Memory: A device or medium used to store information in such a form that it can be understood by the computer hardware.

Memory capacity: The number of actions that a robot can perform in a program or the number of storage locations available.

Memory cycle time: The minimum time interval between two successive data accesses from a memory.

Memory, random access: Also referred to as RAM, a memory whose information media are organized into discrete locations or sectors, for example, each of which is uniquely identified by an address, so that data may be recalled from the memory by specifying the appropriate address.

Menu: A display of options on a device, such as a CRT, for user prompting and selection.

Message: A group of words that transports an item of information.

Micro array computer: A special-purpose, multiprocessor system designed for high-speed calculations with arrays of data.

Microprocessor: A single integrated circuit containing most of the elements of a computer.

Microsecond: One-millionth of a second.

Modem: A contraction of modulator-demodulator, this term can mean either that a modulator and demodulator of a modem are associated at the same end of a circuit, or the modulator and the demodulator of a modem are associated at opposite ends of a circuit to form a channel.

Modular robot: A robot made by assembling pre-engineered modules.

Monitor: An operating programming system that provides a uniform method for handling the real-time aspects of program timing, such as scheduling and basic I/O functions.

Multiplex: The transmission of multiple data bits through a single transmission line by means of a "sharing" technique.

Network: The interconnection of a number of devices by data communications facilities. Local networking is the communication network internal to a robot. Global networking is the ability to provide communications connections outside the robot's internal system.

Noise: An extraneous signal in an electrical circuit that is capable of interfering with the desired signal; also, any disturbance that interferes with the normal operation of a device or system; also, any unwanted information in a digitized image.

Numerical control: The control of machine tools by numerical devices.

Octal: Pertaining to a base 8 numbering system, with the digits 0, 1, 2, 3, 4, 5, 6, and 7 representing all quantities.

Off-line programming: Computer programming development on a system separate from the computer on board a robot, as distinguished from on-line programming, which is a computer program development on the system included in a robot.

Offset: The count value output from an A/D converter resulting from a zero input analog voltage. Used to correct subsequent nonzero measurements.

Open loop: A system without feedback.

Operating range: The reach capability of a robot; also, the work envelope of a robot.

Operating system: A group of programming systems operating under the control of a data-processing monitor program.

Operation, parallel: Operates on all bits of a word simultaneously.

Operation, serial: The flow of information through a computer in time sequence, usually by bit but sometimes by characters.

Output: Information transferred from the internal storage of a computer to an output device or external storage.

Parameter: In a subroutine, a quantity that may be given different values when the subroutine is used in different main routines or in different parts of one main routine, but that usually remains unchanged throughout any one such use. In a generator, a quantity used to specify I/O devices, to designate subroutines to be included, or to describe the desired routine to be generated.

Parity bit: A binary digit appended to an array of bits to make the sum of all the bits always odd or always even.

Parity check: A check that tests whether the number of ones or zeros in an array of binary digits is odd or even.

Part classification: The identification of different parts by a robot, usually by means of a vision system.

Patch: A section of coding inserted into a routine to correct a mistake or to alter the routine. Explicitly, it transfers control from a routine to a section of coding and back again.

Payload: The maximum weight that can be carried by a robot at normal speed; also called workload.

Peripheral: I/O equipment used to make hard copies or to read in data from hard copies.

Pick-and-place robot: A simple robot with two to four axes of motion and little or no trajectory control.

Pitch: The up-and-down motion at an axis or rotation about a horizontal axis when the arm is horizontal.

Pixel: A picture element, that is, an element of the matrix of gray-scale values that is assumed to be a uniform shade of gray in creating the digital image. The more pixels processed per second, the faster the robot can recognize objects. The smaller the pixel size, the less chance there is the robot will make a mistake.

Point operation: Transformation of the gray-scale value of the pixels according to given function, such as thresholding or contrast enhancement.

Point-to-point robot: A servo or nonservo driven robot with a control system for programming a series of points without regard for coordination of axes, in which the intermediate points (and often the rate of motion) are not controlled.

Program: A plan for the solution of a problem, including plans for the transcription of data, coding for the computer, and the absorption of results into the system. The list of coded instructions is called a routine; also, to plan a computation or process from the asking of a question to delivery of results, including the integration of the operation into an existing system.

PROM: An acronym for programmable read-only memory, a memory that can be changed or initialized more than once by the user.

Pseudo (OP) instruction: A symbolic representation of information to a compiler or interpreter. A group of characters having the same general form as a computer instruction, but not executed by the computer as an actual instruction.

RAM: Random access memory.

Range: A characterization of a variable or function; also, all the values that a function may possess.

Rated load capacity: The amount of weight a robot is capable of lifting reduced by a factor of safety.

Read: To copy, usually from one form of storage to another; also, to sense the meaning of arrangements of hardware; also, to sense the presence of information in a recording medium.

Reader: A device that senses information stored in an off-line memory medium and generates equivalent information in an on-line memory device.

Real time: Refers to the actual time during which a physical process or operation takes place; also, the time logged on a robot, in which it is or could be performing work.

Register: A memory device capable of containing one or more computer bits or words.

Repeatability: A measurement of the deviation between a taught location point and the played-back location, under identical conditions of load and velocity; also, the repetitive accuracy or closeness of agreement of repeated position movement under the same conditions at the same location.

Resolution: The number of bits by which the gray-scale values of a digital image is represented; also, the smallest distance increment that can be commanded by a robot controller.

Robomation: A contraction of "robot" and "automation," it refers to the use of robots to control and operate equipment or machines automatically.

Robot: A reprogrammable, multifunctional manipulator designed to move parts, materials, tools, or specialized devices through various programmed motions for the performance of a variety of tasks. In nonindustrial applications the term may be applied to any mechanical device that is able to perform some task or movement under automatic control.

Robotics: The study of robots; also, the science of designing, building, and applying robots.

Roll: Circular motion at an axis, a rotation about the link axis.

ROM: Read-only memory. A digital memory containing a fixed pattern of bits, generally unalterable by the user.

Routine: A series of computer instructions that perform a specific, limited task.

RS232: Standard computer interface data link protocol used for computer communications.

Scan: To examine signals or data point by point in a logical sequence.

Sensor: A transducer or device that transmits information to a robot's controller.

Serial interface: A method of data transmission that permits transmitting a single bit at a time through a single line, used where high-speed input is not needed.

Servo mechanism: A closed-loop control system for a robot in which the computer issues commands, the motor drives the arm, and a sensor measures the motion and signals the amount of the motion back to the computer repeatedly until the arm is repositioned to the point requested; as distinguished from nonservo mechanism, which refers to a closed-loop control system in which the motion of the robot arm is controlled by mechanical stops, usually the beginning and end points of a path of motion.

Servomotor: A motor used to control position. Usually, servomotors are electric, but there are hydraulic and pneumatic ones as well.

Setpoint: The required or ideal value of a controlled variable, usually preset in the computer or system controller by an operator.

Shift: To move information serially right or left in a register of a computer.

Signal processing: Complex analysis of waveforms to extract information.

Significant digit: A digit that contributes to the precision of a numeral. The number of significant digits is counted beginning with the digit contributing the most value and ending with the digit contributing the least value, called the least significant digit.

Simulator: A device or computer program that performs simulation.

Smart sensor: A sensor whose output depends on internal data or on input from another part of the system.

Software: The programs, routines, or supporting documentation that instruct the operations of a computer.

Source language: The symbolic language comprised of statements and formulas used to specify computer processing. It is translated into object language by an assembler, interpreter or compiler, and is more powerful than an assembly language because it translates one statement into many items.

Spherical coordinates: Spatial coordinates defined by two angles and a distance.

Statement: A meaningful expression or generalized instruction in a source language.

Stop: A mechanical constraint on motion.

Subroutine: A series of computer instructions to perform a specific task for many other routines. It is distinguished from the main routine because it requires as one of its parameters a location specifying where to return to in the main program after its function has been accomplished.

Symbolic coding: Any coding system in which symbols rather than actual machine operations and addresses are used.

Syntax: The structure of expressions in a language; also, the rules governing the structure of a language.

System: A collection of parts or devices that forms and operates as an organized whole through some form of regulated interaction.

Table look up: The procedure for obtaining the function value corresponding to an argument from a table of function values.

Teach pendant: The control box that an operator uses to guide a robot through the motion of a desired task. The motions are recorded by the control memory of the robot for future playback in the performance of the task.

Teleoperator: A mobile, robotlike device controlled by a human. Teleoperators are generally used in hazardous environments.

Thresholding operation: An operation in which values are compared to a fixed threshold value and the outcome of the comparison used to set an output value.

Tool: A term used loosely to define an instrument attached or mounted to the end of a robot arm, such as a robot arc-welding torch.

Transducer: A device that converts energy from one form to another.
Transfer vector: A transfer table used to communicate between two or more programs.
TTY: Teletypewriter.
T3: The Tomorrow Tool, one of Cincinnati Milacron's robots.
VAL: The Unimate programming language.
Variable: A quantity that can assume any of a given set of values.
Vision system: A device that can collect data and form an image that can be interpreted by a robot computer to determine an appropriate position or to "see" an object.
Word: A set of bits comprising the smallest addressable unit of information in a programmable memory.
Word length: The number of bits in a word.
Work cell: A manufacturing unit consisting of one or more work stations.
Work envelope: The outline surface of a robot's work volume.
Work station: A manufacturing unit consisting of one robot and the machine tools, conveyors, and other equipment with which it interacts.
World coordinates: A coordinate system based on the plant floor.
Wrist: The manipulator device located between the robot arm and end effector. A general-purpose wrist has 3 degrees of freedom.
Write: To deliver data to a medium.
Yaw: Side-to-side motion at an axis. Rotation of the wrist about a vertical axis when the arm is extended horizontally.
Zero point: The origin of a coordinate system.

Index

Active cord mechanism, 31
Active imaging, 118, 123
Actuator, 31, 54
AL program, 103
Albus, James S., 120
ALGOL, 81
Ambient lighting, 110, 113
AMF, 19
AML program, 102
Aperture control, 113
Applicon, 121
Artificial intelligence language, 81
Androbot, 217
Anthromorphic, 27
Architectures, 116
Arc welding, 152
Arc-welding torch, 38
Argonne National Laboratory, 12
ARMBASIC, 83
Armstrong, Neil, 13
Artificial intelligence, 5, 67
Assembly language, 82
ASCII, 72
ASEA, 206
ASEA robot, 153
ASEE, 190
ASME, 190
Asimov, Isaac, 1, 217
Assembly tools, 40
Astez Corporation, 131
Automata, 10
Automated factories, 214

Automatic inspection, 106
Automatic puppets, 10
Automated machinery, 2
Autonomous robots, 133
Ayres, R. U., 18

Babbage, Charles, 10
Babbitt, Steward S., 10
Bakaleinikoff, Bill, 217
Ball screw, 31
Bardeen, John, 18
Barry Wright Corporation, 132
BASIC, 81
Basic industries, 139
Basic producer, 140
Battele Pacific Northwestern Laboratories, 77
Bellows, 38
Bell Telephone Laboratories, 18
Bend, 32
Bendix, 210
Bergland, 13
Berlichingen, Gotz von, 11
Binary images, 123
Binary numbers, 81
Binocular vision, 106
Bin picking, 165
Birmingham, England, 80
Blocks world, 111
Bonner, Susan, 83
Boorstin, D. J., 1, 9, 66
Boston arm, 20

249

BRA, 20
Brains on Board, 133
Bratton, Walter, 18
Brigham Young University, 121
Brightness sensitivity, 107

CAD, 83
CAM, 83
Cambridge University, 18
Capek, Karel, 1
Cathode ray tube, 97
Carnegie-Mellon University, 133, 226
Cartesian, 24
Cartwright, Edmund, 10
Case Western Reserve University, 132
Casting process, 148
Cetron, M. J., 187, 200
Charles Stark Draper Laboratory, 44
Chevrolet Motor Division of GM, 126
Christenson, Hank, 121
Christiansen, D., 200
Chrysler Newark, 151
Cincinnati Milacron, 20
CINTURN, 182
Closed loop system, 47
Close path operation, 97
COBAL, 81
Collision avoidance, 134
Color vision, 106
Colt, Samuel, 10
Common human senses, 73
Common sense, 1, 68
Compensation network, 54
Compiler program, 82
Compliant end effector, 44
Computer aided design, 69
Computer vision, 110
Computer Vision Corporation, 121
ComRo, 134
ComRo I, 220
ComRo ToT, 220
CONSIGHT, 118
Contact sensors, 130
Continuous path motion, 56
Control Automation, 126
Controlled path motion, 58
Controller, 24
Converter industry, 140

Conveyors, 177
Coordinated motion, 58
Copperweld Robotics, 125
Cubot, 75
Curved surfaces, 121
Cycle start point, 97
Cyclindrical coordinate robot, 25
Cyclindrical lens, 113

DC-2, 218
Decisions, 68
Degree of freedom, 25
Descartes, Rene, 24
Department of Energy, 13
Deburring, 156
Denby, 219
DeVilbiss/Trallfa, 160
Devol, George, 18
Digital differential analyzer, 86
Digitizing, 116
Direct drive, 34
Domestic robots, 216
Dorf, R. C., 221
Drilling, 146

Eaton Corporation, 52
Economic Recovery Tax Act, 203
Edinburgh arm, 19
Editor program, 82
EDSAC, 18
Egyptian water clock, 10
Einstein, Albert, 68
Electric power unit, 47
Electromagnetic gripper, 38
End effector, 36
England, 205
Engelberger, J. F., 106, 217
ENIAC, 18
Ernst, H. A., 19
Error signal, 54
Everett, Bart, 133, 217
Expert system, 68
External sensor, 108

Fabricator industry, 140
Featureless curved objects, 118

Feedback, 47
Fish eye lens, 111
Fitzgerald, C. T., 189
Focus control, 113
Force-torque sensor, 131
Force sensor, 110
Forming operations, 146
FORTH, 137
FORTRAN, 81
Fractals, 112
Frame grabber, 117
France, 205, 206
Friction gripper, 36
Fukui, I., 137

GCA Corp., 210
Gear drives, 31
Geduld, H. M., 10
General Electric, 19, 121, 210
General Electric Cicero plant, 156
General Electric TN 2500 camera, 126
General Motors Technical Center, 118
Gimbal mounting, 137
Global positioning system, 134
Goertz, Ray C., 12
Gore, Alan, 187, 192
Gottesman, R., 10
Gray levels, 111
Gripper, 36
Groover, M. P., 139
Guterl, F., 193

Hall, C., 209
Hall, E. L., 75
Harmon, Leon D., 132
Harrell, R. C., 225
Harvard University, 18
Hearing, 107
Heat treatment, 149
Heer, Ewald, 13
HERO I, 82, 134
Herodotus, 11
Hexadecimal, 82
Hierarchical architecture, 137
Hierarchical control, 60
Hierarchical tree, 111
Histogram, 117
Hitachi robot, 153

Holland, Stephen, 118
Holmes, J. G., 182
Home position, 97
Howard, J. M., 173
Human brain storage capacity, 72
Human factor engineers, 194
Human factor experts, 194
Human-robot labor costs, 170
Hydraulic power unit, 44

IBM, 18, 210
IEEE, 190
IIE, 190
Illumination system, 113
Image processing, 110
Image synthesis, 122
Impact wrench, 40
Index registers, 88
Industrial renaissance, 8
Industrial revolution, 10
Intel Corporation, 19
Intelligence, 67
Intelligent robot, 5
Internal rate of return, 181
International Harvestor, 149
International Robotics, 219
Interpreter language, 82
Isaac, G. W., 224

Jacquet-Droz, Pierre and Henri-Louis, 10
Jacquard, Joseph Marie, 10
Japan, 19
JIRA, 20, 205
Jointed arm robot, 27
Joint angle coordinates, 58
Joint Economic Committee, 203
Joint interpolated motion, 58
JPL, 13

Kawasaki Heavy Industries, 19
Kawasaki Laboratories, 44
Kelley, Robert, 165
Kemle, J., 209
Kinesthetic sense, 108
Kutcher, M., 143

Labor unions, 199
Ladder logic, 82

Ladle, 38
Laser beam machining, 147
Laser range finder, 129
Lathe, 10
Learning, 70
Leg-type robot, 32
Line projection, 118
Line tracking, 61
Linear actuator, 38
LISP, 81
LOGO, 222
LORAN system, 134
Lord Corporation, 131
Luddites, 199

Machine intelligence, 67
Machine Intelligence Corporation, 123
Machine language, 81
Machining process, 146
Maillardet, Henri, 10
Mainline sequence, 97
Malone, Robert, 10
Manhattan project, 13
Manipulator, 3
Manufacturing, 139
Manufacturing industries, 140
Manufacturing language, 81
Manufacturing processes, 139
Master-slave manipulator, 12
Maudslay, Henry, 10
MAZAK plant, 214
McLean, C., 145
MERV, 133
Microbot, 33, 83
MIG welding, 153
Miller, S. M., 18
Milling, 146
Milling machine, 10
Minsky, Marvin, 71
MIT, 19, 71
Mobile robot, 31
Modular robot, 62
Moore, Gordon, 19
Moravec, Hans, 133
MOVIE.BYU, 121
Multifingered hand, 38
Multifingered manipulator, 73
Myoelectric, 11

NASA, 13, 73
National Bureau of Standards, 81, 120
Natural language, 70
Naval Postgraduate School, 133, 217
Nissan factory, 2
Noncontact sensor, 108
Nonservo robot operation, 52
Nonservo system, 47
Nordson robot, 160
Noyce, Robert, 19

Oak Ridge National Laboratory, 16
Octal digits, 81
Odetics, Inc., 227
ODP2, 218
Off-line programming, 81, 83, 104
Ohio State University, 226
Open loop system, 47
Operands, 88
Operating system, 82
OSHA, 176
OTA, 191

Pan control, 113
PASCAL, 81
Passive imaging, 110
Payback period, 180
PEGASUS, 135
Photoelectric cell, 54
Piezoelectric transducer, 131
Piston and cylinder, 31
Pitch, 32
Planet Corporation, 18
Plasma arc cutting, 147
Pneumatic power unit, 47
Point-to-point operation, 56
Polarizing filters, 113
Polaroid, 108
Pollard, Willard L. V., 11
Polygonal surfaces, 118
Position feedback, 54
Power source, 24, 44
Prab Robot, 52
Prab Robots Inc., 162
Process industries, 140
Process tooling, 38
Production, 139

Production control language, 81
Programmable controller, 44, 82
Programming a robot, 80
Program counter, 88
Prosthetics, 11
Proximity sensor, 127
Proximity sensors, 134
Psychologists, 194

Race mode, 58
Radioactive materials, 13
Range sensor, 127
Raphael, B., 133
RB5X, 134, 220
Rectangular, 24
Rectilinear, 24
Relevant information, 75
Reich, F. R., 77
Remote center compliant device, 44
Remote control, 135
Remote drive, 34
Remote manipulator, 7
Renault, 34
Repeatibility, 55
Resolution, 56
RIA, 20, 140, 168
RI/SME, 191
ROBART, 133, 217
Robertson, I., 195
Robot, 1
Robot citrus harvestor, 224
Robot definition, 2
Robot Louis Stevenson, 219
Robot Redford, 218
Robot sheep shearer, 225
Robot Times, 192
Robot vision, 106
Robot Vision Systems Inc., 126
Robot Wars, 220
Robot X News, 206
Robotics, 1
Roehampton arm, 19
Roessing, W., 219
Roll, 19
Roselund, Harrold A., 11
Rubik's cube®, 75
Russia, 205

Sagan, Carl, 72
Sampling theorem, 121
Scott Instruments, 80
Sensor integration, 138
Sequencer, 52
Sergius, Marcus, 11
Servo, 49
Servo robot operation, 54
Servo system, 54
Shaft encoder, 106
Shakey robot, 133
Shannon, Claude, 18
Shaping, 146
Shin, K. G., 83
Shockley, William, 18
Shunk, D. L.,
Shutter mechanism, 137
Shuttle system, 62
SICO, 219
Silhouette image, 112
Sitkins, F. Z., 188
Skinner hand, 38
SME, 20, 191
Smell, 107
Smith, Donald, 192
Smith, Roger, 209
Snowflake curve, 112
Sociologists, 194
Solid-state camera, 115
Southern California Android Amusement Co., 218
Spatial acuity, 107
Spatial array, 108
Spherical coordinate robot, 27
Spot welding, 150
Spot-welding gun, 38
Spray painting, 157
Spray-painting gun, 38
Square D, 209
SRI International, 123
Stack pointer, 88
Stackhouse, Ted, 32
Standardization, 214
Stanford arm, 20
Stanford University, 226
Stationary base line tracking, 62
Stauffer, R. N., 173
Stepper motors, 83

Stereobot, 220
Strobe lighting, 113
Suction gripper, 36
Superior Robotics, 217
Surface description, 121
Surface recognition, 121
Surface representation, 122
Surface shading, 123
Susnjara, Ken, 174
Sweden, 205

Tachometer, 54
Tactile, 5
Tactile sensor, 129
Tactility, 107
Tarvin, Ronald, 176
Teach-by-example programming, 83
Teach pendant, 55, 73
Teleoperator, 13
Thalidomide, 11
T3 program, 94
T3, the tomorrow tool, 20
Three-dimensional surface, 111
Three roll wrist, 35
Tilt control, 31
Tokyo University, 31
Tomy Corp., 168
Torque sensor, 110
Touch, 106
Training center, 190
Transmission, 34
Trevelyan, J. P., 225
Truxal, C., 198
Turning, 146
Turtle, 168

Ultrasonic sensor, 108
Umetani, Y., 31
Underemployment, 194
Unimation, 18
Universal machine, 9
University of Cincinnati, 226
University of Florida, 224

University of Illinois, 78
University of Michigan, 195
University of Pennsylvania, 18
University of Rhode Island, 165
University of Tennessee, 75
University of Western Australia, 225
Urbanization, 196
U.S. Army, 19
U.S. Robots, 209
USSR, 201

VAL program, 100
Vedder, R. K., 205, 208
Velocity feedback, 54
Venture capitalists, 209
Verbex, 80
Vestibular sense, 108
Vidicon, 115
Viking Lander, 221
Voltan Corp., 80

Wabot I, 220
Watt, James, 10
Weisel, W. E., 189
Weld pool, 153
Welding, 150
West Germany, 205
Westinghouse, 209
Wheel slippage, 136
Whitney, Eli, 10
Wolkomir, R., 216
Work envelope, 25
Work volume, 25
World coordinates, 58
World of Robots Corp., 219
Wright, J., 193
Wrist, 32
Wrist faceplate, 97
Wrists, 7

Yaw, 32

Zoom control, 113